SpringerBriefs in Molecular Science

Electrical and Magnetic Properties of Atoms, Molecules, and Clusters

Series editor

George Maroulis, Patras, Greece

W0230283

More information about this series at http://www.springer.com/series/11647

Vladimir Goncharov

Non-Linear Optical Response in Atoms, Molecules and Clusters

An Explicit Time Dependent Density Functional Approach

 Springer

Vladimir Goncharov
Physics and Astronomy
Vanderbilt University
Nashville, MA
USA

ISSN 2191-5407   ISSN 2191-5415 (electronic)
ISBN 978-3-319-08319-3 ISBN 978-3-319-08320-9 (eBook)
DOI 10.1007/978-3-319-08320-9

Library of Congress Control Number: 2014945342

Springer Cham Heidelberg New York Dordrecht London

Printed on acid-free paper

Springer is part of Springer Science+Business Media (www.springer.com)

Contents

Acronyms

ALDA	Adiabatic Local Density Approximation
CG	Conjugate Gradients minimization procedure
DFPT	Density Functional Perturbation Theory
DFT	Density Functional Theory
EFISH	Electric Field-Induced Second Harmonic
EOKE	Electro Optic Kerr Effect
EOS	Electro-Optical Switch
FPS	Full Permutation Symmetry
IDRI	Intensity Dependent Refractive Index
IPS	Intrinsic Permutation Symmetry
LB94	van Leeuwen-Baerends exchange-correlation functional
LDA	Local Density Approximation
OR	Optical Rectification
PZ	Perdew-Zunger exchange-correlation functional
RPA	Random Phase Approximation
RT-TDDFT	Real-Time Time-Dependent Density Functional Theory
SCF	Self Consistent Field
SHG	Second Harmonic Generation
SOS	Sum Over States
TDDFT	Time-Dependent Density Functional Theory
THG	Third Harmonic Generation
TPA	Two Photon Absorption
VWN	Vosko-Wilk-Nusair exchange-correlation functional

Symbols

A	Vector potential
D	Degeneracy factor
E	Electric field
$\hat{H}$	Hamiltonian operator
n	Refractive index
P	Polarization
$\hat{U}$	Evolution operator
V_H	Hartree potential
V_{XC}	Exchange-correlation potential

Greek Symbols

α	Polarizability
β	First Hyperolarizability
γ	Second Hyperolarizability
ϕ	Single particle wavefunction
$\hat{\psi}$	Destruction Field operator
$\hat{\psi}^{+}$	Creation Field operator
$\hbar$	Plank's constant
ρ	Electronic Density
$\chi^{(n)}$	nth Order Nonlinear Susceptibility
ω	Frequency
Ω	Volume of unit cell

Chapter 1
Introduction

1.1 The Need for Highly Nonlinear Optical Materials

We rarely describe objects or phenomena as "non-linear" when speaking of every-day life. In conversational language the word "non-linear" can be associated with things that are out of the ordinary (sometimes alarmingly so). Yet the real world is not linear. The finiteness of most everything that surrounds us—such as objects, forces, space—makes our world essentially non-linear. One of the simplest non-linear objects is a switch. One push, and it goes from closed state to open, letting a powerful current through. An electro-optical switch (EOS) is not much different. A change in electric potentials, and the crystal becomes transparent, letting through a beam of light. A reverse change in potential closes the pathway for the beam of light. These electro-optical switches are expected to be the future elementary base of telecommunication and computing devices. The only problem, or more precisely the major problem with EOS, is that we do not have materials that could be used to make an EOS on a micron scale or smaller [1–8]. Yet smaller it must be, because the existing device, the purely electronic switch, is easily manufactured on a sub-micron scale. The lack of known materials that are sufficiently non-linear is aggravated by the lack of an established theory (besides a few guiding principles) on how to in-crease non-linearity. The search for advanced electro-optical materials is therefore concentrated in the hands of experimentalists. The experiments at certain point in-variably involve a laser and a sample of new material, frequently synthesized and characterized at a high cost using time-consuming procedures. Then follow hours of data analysis by a team of scientists; this analysis typically leads to another ex-periment. The computational methods presented in this brief allow us to simulate light-matter interaction experiments as close as possible, but with no synthetic costs and with minimal time costs. Thanks to the maturing field of supercomputing, it is possible to run hundreds of these virtual experiments within a short time.

To put the problem on a quantitative footing we need a closer look at what the requirements are for optical processors and how they translate into the desirable

© The Author(s) 2014
V. Goncharov, *Non-Linear Optical Response in Atoms, Molecules and Clusters*,
SpringerBriefs in Electrical and Magnetic Properties of Atoms, Molecules, and Clusters,
DOI 10.1007/978-3-319-08320-9_1

properties of nonlinear materials. The main requirements for the new generation of optical processors are [9, 10]:

- Reduced operational voltage for reduced energy per bit processed.
- Reduced size to enable "on the chip" integration with conventional electronic processors.
- Ultra fast response to enable terahertz and all-optical processing with characteristic response time $\tau \sim 10^{-13}$ s.

Although optical modulator designs can vary, mechanism of a typical optical modulator depends on a phase shift $\Delta\phi$ which is related to induced change in refractive index $\Delta n(\lambda)$ (taken at wavelength λ) and the length of active wave guide L, where interaction occurs:

$$\Delta\phi = \frac{\pi \, \Delta n(\lambda) L}{\lambda}. \tag{1.1}$$

Change in refractive index is proportional to nonlinearity (nonlinear refractive index) n_2 and square of electric field $|E|^2$: $\Delta n = 2n_2|E|^2$ [11]. Another, more general way of accessing nonlinear characteristic of optical switch is the nonlinear waveguide . parameter γ_A [12–14]:

$$\gamma_A = \frac{2\pi n_2}{\lambda A}, \tag{1.2}$$

where A is footprint.[1] Greater efficiency to which larger numerical values of γ_A correspond require larger nonlinearity and smaller size. It naturally follows from the above that increasing nonlinearity is the direct way of reducing the size of optical processor and improving its effectiveness. At the same time, increasing nonlinearity of material normally leads to reducing operational electric field because nonlinear response is proportional to the nonlinearity and the square of electric field. Thus, two key characteristics are directly related to enhanced nonlinearity of optical material. The requirement of ultra-fast response could be superfluously satisfied if the chief mechanism of nonlinear response of the device is purely electronic, because the characteristic response time of the device is on the order of a femto second $\tau_{el.} \sim 10^{-15}$ [11]. Unfortunately, electronic nonlinear response in ordinary materials is relatively weak, in fact weaker by two orders of magnitude than response involving molecular re-orientation [11].

Thus, achieving all three key requirements for the new generation of optical processors is not a trivial task. The need for new nonlinear optical materials could be further illuminated if we compare nonlinear coefficients of conventional materials with the most advanced specimens and with characteristics actually needed for a hypothetical optical processor that could compete with modern integrated circuit. Let us assume that an all optical device could be constructed with characteristic size $L = 325$ nm, operating at $\lambda = L$ and pump intensity $I \sim 10^{12}$ (W/m^2). This will imply electric field strength $|E| \sim 10^{-3}$ (V/Å) and corresponding potential of ~ 1 V.

---

[1] Here, footprint is defined as an effective area where the interaction between photons and material occurs.

Modulation with maximum phase shift of $\Delta\phi = \frac{\pi}{2}$ will require change in refractive index $\Delta n \sim 1$, and corresponds to optical nonlinearity $n_2 \sim 10^{-12}$ (m^2/W). Conventional materials have n_2 nonlinearity within 10^{-23}–10^{-20} (m^2/W). For example ordinary air has $n_2 = 5.0 \times 10^{-23}$ (m^2/W) and Al$_2$O$_3$ has $n_2 = 2.9 \times 10^{-20}$ (m^2/W). Specialized optical materials such as SF-59 glass (Schott) and As$_2$S$_3$ glass have $n_2 = 3.3 \times 10^{-19}$ (m^2/W) and 3.0×10^{-17} (m^2/W) respectively. However, the highest n_2 currently measured in nanostructured plasmonic (meta)materials, such as patterned gold nanoparticles in glass is 2.6×10^{-14} (m^2/W) [11]. Therefore, there is both the need and the room to increase the nonlinearity of optical materials by factor of $\times 10^2$ to $\times 10^5$.[2]

The work presented here is aimed at developing new computational framework that can simplify and speed up the search for new nonlinear optical materials, and facilitate study of light-matter interactions from quantum mechanical principles. The work currently addresses two aspects.

One is realistic modeling of interaction of light and material. This is achieved by constructing a molecular structure and then applying a model laser pulse. The system is then evolved in real time by solving time-dependent Schrödinger equation.

The other aspect is extraction of response functions from these simulations. Most of the nonlinear optical phenomena, such as Kerr effect, Two Photon Absorption, optical rectification , Intensity Dependent Refractive Index and Second and Third Harmonic Generation have corresponding response functions. Optical response functions have many names: susceptibilities, polarizabilities, both linear and nonlinear, and hyperpolarizabilities. They provide a quantitative measure for the response of matter to electromagnetic field, enabling evaluation of material under study for prospective applications. If response function is known, it can be used to model relevant processes that go beyond initial "experiment" from which it was obtained. In the next section we introduce basic terminology and highlight utility of the optical response functions.

1.2 Nonlinear Optical Phenomena in Terms of Response Functions

The equation describing propagation of a monochromatic wave

$$\mathbf{E}(\mathbf{r}, t) = \mathbf{E}(\mathbf{r})e^{i\omega t} + c.c. \tag{1.3}$$

---

[2] We have not discussed second order nonlinear materials that are utilized in Pockels effect based devices primarily because all-optical processing is not feasible in them. However, the situation with these materials is similar to what has been described: there is two to five orders of magnitude gap in nonlinearity that needs to be crossed in order to satisfy requirements of modern optoelectronics applications. In addition, the second order materials with the highest nonlinearity are organic polymers [6] that overwhelmingly suffer from thermal stability problems [15].

with wave number $k^2(\omega) = \frac{n(\omega)\omega}{c}$ in media with refractive index $n(\omega)$ and a time dependent source

$$\mathbf{P}^{(m)}(\mathbf{r}, t) = \mathbf{P}^{(m)}(\omega)e^{\iota\omega t - i\mathbf{k}\times\mathbf{r}} + c.c. \tag{1.4}$$

could be written as:

$$\nabla^2\mathbf{E}(\mathbf{r}) + k^2(\omega)\mathbf{E}(\mathbf{r}) = -4\pi\frac{\omega^2}{c^2}\mathbf{P}^{(m)}(\omega)e^{-i\mathbf{k}\times\mathbf{r}}. \tag{1.5}$$

Let the source be mth term in decomposition of total polarization in powers of electric field $\mathbf{E}'$:

$$\mathbf{P}^{total}(\omega) = \sum_{i=1}^{N_{max}} \mathbf{P}^{(i)}(\omega, (\mathbf{E}')^i). \tag{1.6}$$

Then, Eq. (1.5) will be related to propagation of nonlinear polarization wave of mth order. Its solution with appropriate boundary conditions gives mathematical description of specific nonlinear process of mth order. The source term is the nonlinear polarization, which is expressed in terms of response functions. For monochromatic electric fields it could be written as:

$$P_i^{(m)} = \sum_{j...k} \chi_{ij...k}^{(m)} E_j \ldots E_k. \tag{1.7}$$

The coefficients $\chi_{ij...k}^{(m)}$ are mth order response functions, or susceptibilities. Their tensor nature plays the key role in determining propagation of nonlinear wave. For example, in case of SHG, which is a second order process, orientational dependence of reflected second harmonic from crystal with $\bar{4}3m$ symmetry is found from [16, 17]

$$P_x^{(2)}(2\omega) = \chi_{xyz}^{(2)} E_y E_z,$$

$$P_y^{(2)}(2\omega) = \chi_{yzx}^{(2)} E_z E_x,$$

$$P_z^{(2)}(2\omega) = \chi_{zxy}^{(2)} E_x E_y.$$

When the electric vector is polarized along [111] crystal axis, all P-components are equal and P is in the same direction as E. When E is along [100], y and z components are absent and P is zero. Similar considirations apply along the [010] direction. For [011] direction, P has x-component only.

Magnitude of some nonlinear effects could be easily evaluated if corresponding susceptibility is known. For example, in case of IDRI, refractive index is calculated from third order susceptibility taken at fundamental frequency ω of propagating wave:

$$n(\omega) = n_0 + \frac{12\pi^2}{n_0^2 c}\chi^{(3)}(\omega)\,I,$$

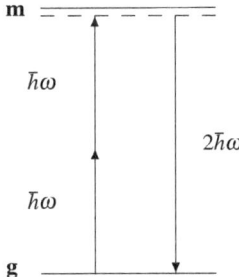

Fig. 1.1 Absorption-emission diagram for Second Harmonic Generation. A molecule in state **g** absorbs two photons, transitions to a virtual state **m** (indicated by *dashed line*), and then emits a photon with energy equal to the energy of the absorbed photons $2\hbar\omega$

where I is intensity of the wave, and n_0 is the linear refractive index.

In short, the main message of this section is that response functions are a compact way of describing key properties of many nonlinear optical processes and materials.

1.3 Brief Account of Nonlinear Optical Processes

1.3.1 Second Harmonic Generation

Second Harmonic Generation[3] is perhaps the most widely known nonlinear effect. In this process two photons combine to produce one photon with double frequency (see Fig. 1.1). The order of a nonlinear process is determined by the number of photons involved. Thus, SHG is a second order process. SHG is an efficient process and is routinely used to upconvert infrared radiation to visible and ultraviolet radiation.

Third Harmonic Generation [19] is analogous to SHG. In this process three photons combine to produce one with triple frequency: $\hbar\omega + \hbar\omega + \hbar\omega \rightarrow \hbar\omega'$.

THG is a much less efficient process than SHG and is typically observed in non-centrosymmetric systems where SHG is suppressed.

1.3.2 Optical Rectification

A second order process in which two photons annihilate simultaneously producing constant polarization is call Optical Rectification. OR was first described by

---

[3] First experiment on second harmonic generation by Franken et al. [18] in 1961 used ruby laser $\lambda = 6940$ Å, a quartz crystal and a pair of filters to detect UV beam at $\lambda = 3470$ Å. The $3J$ laser pulse with duration of a millisecond converted about 10^{11} UV photons per pulse.

Bass et al. [20] who observed that propagation of laser beam through a crystal induces constant polarization at the crystal surface.

1.3.3 Electric Field Induced Second Harmonic

As was already mentioned, SHG is suppressed in molecules and crystals with inversion symmetry.[4] Yet there is a way to generate the second harmonic in this case. The inversion symmetry could be broken with a strong static electric field [21]. The resulting third order process is called Electric Field Induced Second Harmonic (EFISH).

1.3.4 Self-focusing

We have already mentioned Intensity Dependent Refractive Index in Sect. 1.2. IDRI actually encompasses several third order processes, and is described by the real part of third order susceptibility $\chi^{(3)}$. One of the processes that results from dependence of refractive index on the intensity of the propagating wave is self-focusing [22]. If corresponding nonlinear susceptibility $\chi^{(3)}$ is positive, then nonlinear refraction index increases with intensity. When intensity varies in space, as in case of Gaussian beams, the nonlinear medium acts as a positive lens. If intensity is high enough and the medium is large enough the beam may focus to the point of collapse typically leading to extensive material damage. However, the same effect could be used to compensate diffractional de-focusing by adjusting beam intensity. This effect is used in optical fibers to propagate light pulses unperturbed in shape over intercontinental distances.

1.3.5 Two Photon Absorption

While real part of $\chi^{(3)}$ describes IDRI, the imaginary part of $\chi^{(3)}$ is related to Two Photon Absorption [23]. The two photon absorption coefficient is proportional to the intensity of the beam of light, while the linear absorption does not depend on the intensity. In TPA, two photons are absorbed simultaneously by a molecule following by de-excitation with spontaneous emission of several photons at frequencies and directions that generally differ from the initial ones. TPA is a competitive process to THG.

---

[4] See next chapter for details.

1.3.6 Electro-Optic Kerr Effect

Modulation of refractive index n by external electric field when change Δn is quadratic in applied field is called Electro-Optic Kerr Effect [24]. The strength of electric field is typically 10^5 (V/cm). EOKE is the third order nonlinear effect [25] and is directly observable in isotropic fluids such as CS_2 and benzene. EOKE normally requires two optical beams—the pump and the probe. Recently [26] the optical pump was replaced by terahertz electric field to yield all-electrical modulation of optical pulses.

References

1. M.-H. Shih, Nat. Photon. **8**, 171 (2014)
2. A. Facchetti et al., Nat. Mater. **3**, 910 (2004)
3. L. Chen, K. Preston, S. Manipatrini, M. Lipson, Opt. Express **17**, 15248 (2010)
4. M. Paniccia, Nat. Photon. **4**, 498 (2010)
5. H.-T. Chen et al., Nat. Photon. **3**, 148 (2009)
6. Y. Enami et al., Nat. Photon. **1**, 180 (2007)
7. Z.L. Samson et al., Appl. Phys. Lett. **96**, 143105 (2010)
8. C. Videlot-Ackermann et al., Synth. Met. **156**, 154 (2006)
9. G.T. Reed, G. Mashanovich, F.Y. Gardes, D.J. Thomson, Nat. Photon. **4**, 518 (2010)
10. M. Kauranen, A.V. Zayats, Nat. Photon. **6**, 737 (2012)
11. R.W. Boyd, *Nonlinear Optics* (Elsevier, New York, 2008)
12. M. Foster, K. Moll, A. Gaeta, Opt. Express **12**, 2880 (2004)
13. C.C. Koos et al., Opt. Express **15**, 5976 (2007)
14. J. Leuthold, C. Koos, W. Freude, Nat. Photon. **4**, 535 (2010)
15. M. Hayden, Nat. Photon. **1**, 138 (2007)
16. J. Ducuing, N. Bloembergen, Phys. Rev. Lett. **10**, 474 (1963)
17. N. Bloembergen, *Nonlinear Optics* (W. A. Benjamin Inc., New York, 1965)
18. P.A. Franken, A.E. Hill, C.W. Peters, G. Weinreich, Phys. Rev. Lett. **7**, 118 (1961)
19. G.H.C. New, J.F. Ward, Phys. Rev. Lett. **19**, 556 (1967)
20. M. Bass, P.A. Franken, J.F. Ward, G. Weinreich, Phys. Rev. Lett. **9**, 446 (1962)
21. R.W. Terhune, P.D. Maker, C.M. Savage, Phys. Rev. Lett. **8**, 404 (1962)
22. P.L. Kelley, Phys. Rev. Lett. **15**, 1005 (1965)
23. W. Kaiser, C.G.B. Garrett, Phys. Rev. Lett. **7**, 229 (1961)
24. J. Kerr, Phil. Mag. **3**, 321 (1877)
25. P.D. Maker, R.W. Terhune, C.M. Savage, Phys. Rev. Lett. **12**, 507 (1964)
26. M.C. Hoffmann, N.C. Brandt, H.Y. Hwang, K.-L. Yeh, K.A. Nelson, Appl. Phys. Lett. **95**, 231105 (2009)

Chapter 2
Response Functions

2.1 Causal Response

Response functions form a wide class of both classical and quantum quantities. Synonyms of response functions are linear and non-linear susceptibilities of different kinds, as well as polarizability and hyperpolarizabilities. The defining characteristic of a response function $\chi(t)$ is causality of a map that it establishes between perturbing quantity $E(t)$ and a responding quantity $P(t)$.[1] In case of a linear response function, it is accomplished by an integral relation:

$$P(t) = \int\limits_{0}^{\infty} \chi^{(1)}(\tau) E(t - \tau) d\tau. \tag{2.1}$$

One may examine by inspection that (2.1) guarantees that values of $E(t)$ at times earlier than t_0 do not contribute to $P(t_0)$. It also allows for response $P(t_0)$ to persist for all times $t > t_0$ even if field $E(t)$ is zero at these times. For example, taking time profile as delta function for $E(t) = E\,\delta(t)$, and taking $\chi^{(1)}$ as being non-zero only on an interval $0 < t < t_M$ leads to the following response $P(t)$:

$$P(t) = \begin{cases} E\chi^{(1)}(t) \text{ if } 0 < t < t_M \\ 0 \qquad\quad \text{if } t > t_M \end{cases} \tag{2.2}$$

Multiplying (2.1) by $e^{i\omega t}$, integrating in time t from $-\infty$ to ∞, changing variable in left hand side (LHS) $t' = t - \tau$ and using definition of Fourier Transforms one gets frequency domain representation of linear response:

$$P(\omega) = \chi^{(1)}(\omega) E(\omega), \tag{2.3}$$

---

[1] Both $P(t)$ and $E(t)$ are assumed to be observable (i.e. real).

© The Author(s) 2014
V. Goncharov, *Non-Linear Optical Response in Atoms, Molecules and Clusters*,
SpringerBriefs in Electrical and Magnetic Properties of Atoms, Molecules, and Clusters,
DOI 10.1007/978-3-319-08320-9_2

where linear response function is:

$$\chi^{(1)}(\omega) = \int_0^\infty \chi^{(1)}(t)e^{i\omega t}\,dt. \tag{2.4}$$

Sometimes it is convenient to replace (2.4) with a conventional Fourier Transform by multiplying $\chi^{(1)}(t)$ by step function $\theta(t)$ and extending limits of integration to $-\infty$. Generalizations of (2.1) and (2.3) to higher order response are:

$$P^{(n)}(t) = \int_0^\infty \dots \int_0^\infty \chi^{(n)}(\tau_1, \dots, \tau_n)E(t-\tau_1)\dots E(t-\tau_n)d\tau_1\dots d\tau_n, \tag{2.5}$$

$$P^{(n)}(\omega) = \frac{1}{(2\pi)^{(n-1)}} \int_0^\infty \dots \int_0^\infty \chi^{(n)}(\omega; \omega_1, \dots, \omega_n)E(\omega_1)\dots E(\omega_n) \tag{2.6}$$

$$\times\, \delta(\omega - \omega_1 - \dots - \omega_n)\,d\omega_1\dots d\omega_n.$$

Delta function appearing in (2.6) enforces conservation of energy.

2.2 Kramers-Kronig

Causality of response functions leads to several properties that are intrinsic to this class of functions.[2] For linear response, from (2.4) it follows that $\chi^{(1)}(-\omega) = (\chi^{(1)}(\omega))^*$. If ω is complex, then it turns into:

$$\chi^{(1)}(-\omega^*) = (\chi^{(1)})^*(\omega). \tag{2.7}$$

Kramers-Kronig dispersion relations are the consequence of (2.7). Kramers-Kronig relations connect real and imaginary parts of $\chi^{(1)}$ *via* a Hilbert Transform:

$$\Re(\chi^{(1)}(\omega)) = \frac{1}{\pi}\mathscr{P}\int_{-\infty}^\infty \frac{\Im(\chi^{(1)}(\xi))}{\xi-\omega}d\xi, \tag{2.8}$$

$$\Im(\chi^{(1)}(\omega)) = -\frac{1}{\pi}\mathscr{P}\int_{-\infty}^\infty \frac{\mathbb{R}\Re(\chi^{(1)}(\xi))}{\xi-\omega}d\xi. \tag{2.9}$$

---

[2] The proof of these relations could be found in [1].

These are routinely used in calculations as well as in experimental work, where they are used for optical data inversion, for example for deducing dispersion from absorption spectra.

Analog of (2.7) for general nonlinear case exists for real frequencies

$$\chi^{(n)}(-\omega_1, \ldots, -\omega_n) = (\chi^{(n)}(\omega_1, \ldots, \omega_n))^*, \tag{2.10}$$

and for some nonlinear processes for complex frequencies. It has been shown that analogs of Kramers-Kroning (2.8 and 2.9) for nonlinear processes in form of multidimensional Hilbert Transforms do not generally exist. The cases for which they exist include all orders of higher harmonic generation, for which KK takes the following form:

$$\Re(\chi^{(n)}(-n\omega; \omega, \ldots, \omega)) = \frac{1}{\pi} \mathscr{P} \int_{-\infty}^{\infty} \frac{\Im(\chi^{(n)}(-n\omega'; \omega', \ldots, \omega'))}{\omega' - \omega} d\omega', \tag{2.11}$$

$$\Im(\chi^{(n)}(-n\omega; \omega, \ldots, \omega)) = -\frac{1}{\pi} \mathscr{P} \int_{-\infty}^{\infty} \frac{\Re(\chi^{(n)}(-n\omega'; \omega', \ldots, \omega'))}{\omega' - \omega} d\omega'. \tag{2.12}$$

A further discussion of application of Kramers-Kronig relations to nonlinear optics could be found in [2].

2.3 Symmetry Relations

In this section we specialize to the response functions that describe electronic polarization by external electric fields $\mathbf{E}$. These functions are tensors of $(n + 1)$ rank, where n is the order of nonlinearity. Besides symmetry relation (2.10) that follows from causality of response, there are two other kinds of symmetries: one related to structural symmetry of material and another to permutation properties of response function.[3]

2.3.1 Permutation Symmetries

The most general of permutation symmetries is Intrinsic Permutation Symmetry. It follows from the fact that one can not distinguish physical order of the fields appearing in expressions of the following form:

$$\chi_{ij_1 j_2 \ldots j_n}^{(n)}(-\omega_\sigma; \omega_1, \omega_2, \ldots, \omega_n) E_{j_1}(\omega_1) E_{j_2}(\omega_2) \ldots E_{j_n}(\omega_n).$$

---

[3] In depth discussion of this subject could be found in [3].

From this follows property that allows us to permute indices j_k simultaneously with the corresponding frequency ω_k:

$$\chi_{ij_1j_2...j_n}^{(n)}(-\omega_\sigma; \omega_1, \omega_2, \ldots, \omega_n) = \chi_{ij_2j_1...j_n}^{(n)}(-\omega_\sigma; \omega_2, \omega_1, \ldots, \omega_n). \qquad (2.13)$$

As a result, number of distinct permutations enters as a factor for the series of equivalent terms in calculations of polarization response. For example, $\chi_{ijk}^{(2)}(-\omega_\sigma; \omega_1, \omega_2)$ $= \chi_{ik_1}^{(2)}(-\omega_\sigma; \omega_2, \omega_1)$ and second order polarization will become:

$$P_i^{(2)}(-\omega_\sigma) = \frac{1}{2\pi \, 2!} \sum_{jk} D \int \chi_{ijk}^{(2)}(-\omega_\sigma; \omega_1, \omega_2)$$

$$\times \, E_j(\omega_1) E_k(\omega_2) \, \delta(\omega_\sigma - \omega_1 - \omega_2) \, d\omega_2,$$

where $D = 2$ is a number of distinct permutations of fields $E(\omega)$, $2!$—coefficient of Taylor expansion, and 2π is Fourier Transform factor.

In case of lossless media $Im(\chi^{(n)}) = 0$ and IPS becomes Full Permutation Symmetry , where all indices can be permuted simultaneously with corresponding frequencies:

$$\chi_{ij_1j_2...j_n}^{(n)}(-\omega_\sigma; \omega_1, \omega_2, \ldots, \omega_n) = \chi_{j_nj_2j_1...i}^{(n)}(-\omega_n; \omega_2, \omega_1, \ldots, -\omega_\sigma)$$

$$= \chi_{j_1j_2i...j_n}^{(n)}(-\omega_1; \omega_2, -\omega_\sigma, \ldots, \omega_n). \qquad (2.14)$$

In case of lossles $Im(\chi^{(n)}) = 0$ and dispersionless media $Re(\chi^{(n)}) = const.$ one has Kleinman Symmetry which allows one to permute indices without regard to frequencies:

$$\chi_{ij_1j_2...j_n}^{(n)}(-\omega_\sigma; \omega_1, \omega_2, \ldots, \omega_n) = \chi_{j_nj_2j_1...i}^{(n)}(-\omega_\sigma; \omega_1, \omega_2, \ldots, \omega_n)$$

$$= \chi_{j_1j_2i...j_n}^{(n)}(-\omega_\sigma; \omega_1, \omega_2, \ldots, \omega_n). \qquad (2.15)$$

2.3.2 Structural Symmetries

Spatial arrangement of atoms in molecules and solids is frequently symmetric. The point group of material structural symmetry $\mathscr{S}$ is a finite subgroup of the full symmetry group of Hamiltonian. It can be shown that the related response functions must also possess the same point group. Let $S_{nm}^{(g)}$ be a matrix representing gth element of this group. Since a response function of nth order is a tensor of $n + 1$ rank, it transforms according to:

$$\chi'^{(n)}_{p i \dots, j} = \sum_{s k \dots, m} \chi^{(n)}_{s k \dots, m} S^{(g)}_{ps} S^{(g)}_{ik} \dots S^{(g)}_{jm}. \qquad (2.16)$$

If the order of the group is N_g then, there exist N_g constraints of type (2.16) that can be used to reduce the number of independent elements. As an illustration consider inversion symmetry that is an element of T_h, O_h and other point groups. Its matrix representation is $S^{(inversion)}_{ij} = -\delta_{ij}$. In case of even order (nonlinear) response function, from (2.16) follows:

$$\chi'^{(2n)}_{p i \dots, j} = - \sum_{s k \dots, m} \chi^{(2n)}_{s k \dots, m} \delta_{ps} \delta_{ik} \dots \delta_{jm} = -\chi^{(2n)}_{p i \dots, j} = 0. \qquad (2.17)$$

Therefore even orders of nonlinear response vanish if the material possesses inversion symmetry. The tables indicating non-vanishing elements for the first, second and third order response functions for several point groups may be found in [4].

2.4 Quantum Field Theory Response Formalism

Typically, (hyper) polarizabilities are defined as coefficients of Taylor series expansion of polarization $P(t)$:

$$P(t) = \chi^{(1)} E(t) + \frac{1}{2!} \chi^{(2)} E^2(t) + \cdots \qquad (2.18)$$

In this expansion, the hyperpolarizabilities are formally partial derivatives of the "total", generally time dependent polarization in respect to the electric field $E(t)$:

$$\chi^{(2)}_{ijk} = \frac{D \, \partial^2 \, P_i}{2! \, \partial E_j \partial E_k}, \qquad (2.19)$$

where D is degeneracy factor. This definition is purely classical and is frequently supplemented by a qualification such as "…if the series converge, then the hyperpolarizabilities could be defined as (2.18) …". Since convergence of (2.18) generally requires electric field to be small $E << 1$, it raises a question whether the classical definition has any relevance in Nonlinear optics, because in practice the electric field has to be strong in order for the nonlinear phenomena to appear. One may even go as far as to question the existence and the applicability of response functions to the description of any strong field phenomena. In this section we try to address this issue. We define the optical response functions through the quantum density–density response functions that are in turn related to higher order density fluctuations.

To underline the quantum mechanical nature of nonlinear optical response we need to extend the textbook theory of linear response [5] to higher orders. The many-body Hamiltonian is taken in second quantization:

$$\hat{H} = \int d^3x \, \hat{\psi}^\dagger(x) \, T(x) \, \hat{\psi}(x) + \frac{1}{2} \int \int d^3x \, d^3x' \, \hat{\psi}^\dagger(x)\hat{\psi}^\dagger(x')$$

$$\times \frac{e}{|\mathbf{r} - \mathbf{r}'|} \, \hat{\psi}(x')\hat{\psi}(x) + \hat{H}^{ext} \quad (2.20)$$

where $x = (\mathbf{x}, t, spin)$, c_k, $c_k^\dagger$ are field annihilation and creation operators, $\psi_k(x)$ are single particle states and $\hat{\psi}$, $\hat{\psi}^\dagger$ are field operators: $\hat{\psi}(x) = \sum_k \psi_k(x) \, c_k$, $\hat{\psi}^\dagger(x) = \sum_k \psi_k^\dagger(x) \, c_k^\dagger$. The external interaction is described in general by $\hat{H}^{ext} = \int d^3x \, \hat{n}(x) \, \phi^{ext}(x)$, where $\hat{n}(x)$ is density operator $\hat{n}(x) = \hat{\psi}^\dagger(x)\hat{\psi}(x)$. Taking external potential as $\phi^{ext}(x) = e\,\mathbf{r}\cdot\mathbf{E}(t)$ leads to one of the forms of dipole approximation for photon-electron interaction:

$$\hat{H}^{ext} = e \sum_{ij} \langle i | \, \mathbf{r} \cdot \mathbf{E}(t) \, | j \rangle c_i^\dagger c_j = \int d^3x \, \hat{\psi}^\dagger(x) \, e\mathbf{r} \cdot \mathbf{E}(t) \, \hat{\psi}^\dagger(x)$$

$$= e \int d^3x \, \hat{n}(x) \, \mathbf{r} \cdot \mathbf{E}(t). \quad (2.21)$$

Next, we expand the many-body state vector $|\Psi_S(t)\rangle$ in terms of time ordered products of external interaction $T(H^{ext}(t') \dots H^{ext}(t'^{\dots'}))$

$$|\Psi_S(t)\rangle = e^{-\frac{iHt}{\hbar}} \left(1 - \frac{i}{\hbar} \int dt' \, H^{ext}(t')\right.$$

$$\left. - \frac{1}{2!\,\hbar^2} \int dt' \, dt'' \, T(H^{ext}(t')H^{ext}(t'')) + \dots\right) |\Psi_S(0)\rangle \quad (2.22)$$

and use it to compute the density fluctuation $\delta\langle \hat{n}(x)\rangle$:

$$\delta\langle \hat{n}(x)\rangle = \langle\Psi_S(t)|\hat{n}_S(x)|\Psi_S(t)\rangle - \langle\Psi_S(0)|\hat{n}_S(x)|\Psi_S(0)\rangle = \langle\hat{n}(x)\rangle - \langle\hat{n}(x)\rangle_0. \quad (2.23)$$

We observe that the density fluctuation could be represented as a series with kth term being a function of kth power of external potential ϕ^{ext}:

$$\delta\langle \hat{n}(\mathbf{x}, t)\rangle = \sum_k \delta\langle \hat{n}^{(k)}(\mathbf{x}, t; (\phi^{ext})^k)\rangle. \quad (2.24)$$

The non-linear response starts with the second order contribution

$$\delta\langle \hat{n}^{(2)}(x)\rangle = \frac{1}{2!\,\hbar^2} \int d^4x' d^4x'' \, \phi^{ext}(x')\phi^{ext}(x'')$$

$$\times \langle\Psi_S(0)|[[\hat{n}_H(x'), \hat{n}_H(x)], \hat{n}_H(x'')]|\Psi_S(0)\rangle. \quad (2.25)$$

Introducing the second order density-density response function $\Xi^{(2)}$

$$\Xi^{(2)}(x; x', x'') = \theta(t - t')\theta(t' - t'')\frac{\langle\Psi_S(0)|[[\hat{n}_H(x'), \hat{n}_H(x)], \hat{n}_H(x'')]|\Psi_S(0)\rangle}{\hbar^2 \langle\Psi_S(0)|\Psi_S(0)\rangle},$$

(2.26)

the second order density fluctuation could be written as

$$\delta\langle\hat{n}^{(2)}(\mathbf{x}, \omega)\rangle = \frac{1}{4\pi}\int \Xi^{(2)}(\omega; \omega', \omega'', \mathbf{x}, \mathbf{x}', \mathbf{x}'')\phi^{ext}(\omega', \mathbf{x}')\phi^{ext}(\omega'', \mathbf{x}'')$$
$$\times \delta(\omega - \omega' - \omega'')d^3x'd^3x''d\omega'd\omega''.$$

(2.27)

For the finite systems, such as molecules, we can use the density fluctuation to directly compute polarization $\mathbf{P}$ (in practice only a change in polarization $\Delta\mathbf{P}(t)$ is relevant)

$$\mathbf{P} = \int d^3x \, \mathbf{x} \, \delta\langle\hat{n}(x)\rangle,$$

(2.28)

which could also be written as a series analogous to (2.24) :

$$\mathbf{P}(t) = \sum_k \mathbf{P}^{(k)}(t, (E)^k).$$

(2.29)

The second term corresponds to the second order nonlinear optical response:

$$\mathbf{P}^{(2)}(t) = \int d^3x \, \mathbf{x} \, \delta\langle\hat{n}^{(2)}(\mathbf{x}, t)\rangle = \frac{1}{2!}\int \chi^{(2)}_{ijk}(t; t', t'')E_j(t')E_k(t'')dt'dt'', \quad (2.30)$$

where $\chi^{(2)}_{ijk}$ is the first hyperpolarizability. Fourier transforming (2.30) yields

$$P_i^{(2)}(\omega) = \mathscr{K}\int \chi^{(2)}_{ijk}(\omega; \omega', \omega'')E_j(\omega')E_k(\omega'')\delta(\omega - \omega' - \omega'')d\omega'd\omega'', \quad (2.31)$$

where $\mathscr{K}$ is factor from Table 5.2. Comparing (2.27) and (2.31) we see that optical susceptibilities could be obtained directly from density-density response function:

$$\chi^{(2)}(\omega; \omega', \omega'') = \int \Xi^{(2)}(\omega; \omega', \omega'', \mathbf{x}, \mathbf{x}', \mathbf{x}'')\mathbf{x}\mathbf{x}'\mathbf{x}'' d^3xd^3x'd^3x''.$$

The Eqs. (2.29) and (2.18) are both the expansions of the total polarization in the external electric fields, and therefore the terms with the same power of electric field must be equal. This should convince the reader that the hyperpolarizabilities obtained via a classical expansion of the total polarization are in fact quantum mechanical quantities. Their existence and properties are governed by the mechanisms of photon-electron interaction that is specific for a system. Each kth term is related to k-photon process, and the number of the terms is restricted by the energy conservation. Therefore, the classical expansion (2.18) should be viewed as a finite polynomial rather then series, and the question of its convergence is not relevant.

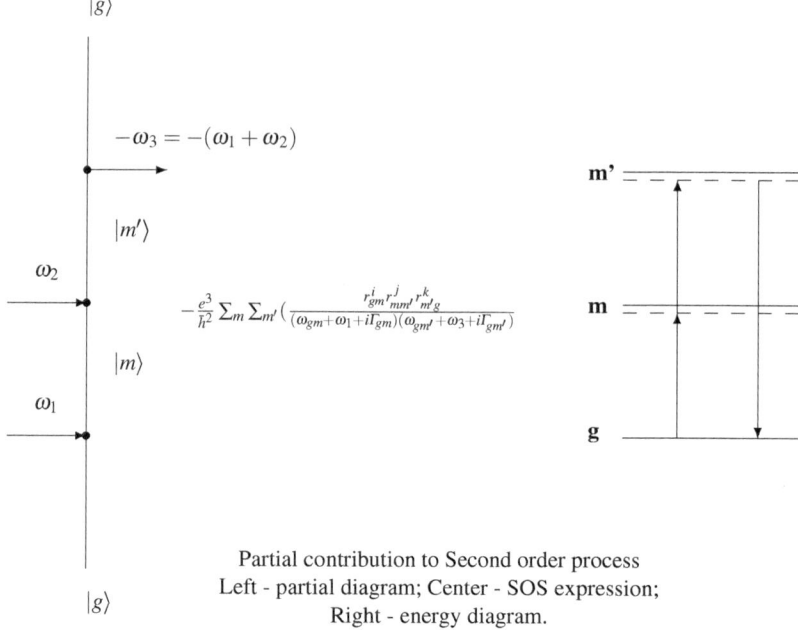

Partial contribution to Second order process
Left - partial diagram; Center - SOS expression;
Right - energy diagram.

Fig. 2.1 A partial diagram for second order process: $\chi^{(2)}(-\omega_3; \omega_1, \omega_2)$. $|m\rangle$ and $|m'\rangle$ are virtual states

2.5 Diagrammatic Technique for Susceptibilities

In this section we present rules that facilitate drawing pictorial representation of nth order of polarization expansions of type (2.31) and writing down corresponding expressions for matrix elements $\chi^{(n)}_{ij...k}$. This diagrammatic technique is analogous to construction of non-relativistic Feynman Diagrams [6, 7]. The resulting expressions for $\chi^{(n)}$ are essentially the same as those one would obtain from matrix elements of electric dipole operator using wavefunctions calculated to nth order of perturbation theory. For nth order process

1. Draw a (vertical) line. On the line draw $n + 1$ vertices.
2. This will partition line into $n + 2$ segments. Label first and the last segments with initial $|g\rangle$ and final states $|g'\rangle$. Label remaining segments with intermediate (generally virtual) states: $|m\rangle$, $|m'\rangle$, ….
3. Each vertex corresponds to a matrix element of external potential, that in case of electric dipole interaction becomes $\langle m' | e\, r_j | m\rangle = e\, r^{j}_{m'm}$. Here r_j is jth Cartesian component of position operator $\hat{r}$. Distribute components over vertices.
4. Draw a (horizontal) arrow in/out of each vertex. Label arrows pointing to vertex with $+\omega$. This corresponds to absorption of photon with energy $\hbar\omega$. Label arrows pointing out of vertex with $-\omega'$. This corresponds to emission of photon with energy $\hbar\omega'$.

5. For each intermediate state $|m\rangle$ write down propagator $\frac{1}{\Delta_{mg}-i\Gamma_{mg}}$ where Δ_{mg} is
   energy of state $|m\rangle$: $\Delta_{mg} = E_m - E_g + \hbar\sum_i \pm\omega_i$, and Γ_{mg} is line width of
   $|m\rangle \rightarrow |g\rangle$ transition
6. Write down expression corresponding to the diagram by summing up over all
   intermediate states m products of $n + 1$ vertices with n propagators.
7. Repeat the steps above for all permutations of frequencies ω_i, sum up resulting
   expressions.

For example, for a second order process that starts at ground state, then absorbs
two photons with energies $\hbar\omega_1$ and $\hbar\omega_2$, then emits a photon with energy $\hbar\omega_3 = \hbar(\omega_1 + \omega_2)$ and ends at ground state one gets diagram (Fig. 2.1) with corresponding
expression:

$$-\frac{e^3}{\hbar^2}\sum_{mm'}\frac{r^i_{gm}r^j_{mm'}r^k_{m'g}}{(\omega_{gm}+\omega_1+i\Gamma_{gm})(\omega_{gm'}+\omega_3+i\Gamma_{gm'})}. \tag{2.32}$$

Diagrams resulting in permutation of ω_1, ω_2 and ω_3 are shown on diagram
(Fig. 2.2), and the summed expression is

$$\chi^{(2)}_{ijk}(-\omega_3;\omega_1,\omega_2) = -\frac{e^3}{\hbar^2}\sum_m\sum_{m'}(\frac{r^i_{gm}r^j_{mm'}r^k_{m'g}}{(\omega_{gm}+\omega_1+i\Gamma_{gm})(\omega_{gm'}+\omega_3+i\Gamma_{gm'})}$$

$$+\frac{r^j_{gm}r^k_{mm'}r^i_{m'g}}{(\omega_{gm}+\omega_1+i\Gamma_{gm})(\omega_{gm'}-\omega_2+i\Gamma_{gm'})}$$

$$+\frac{r^k_{gm}r^j_{mm'}r^i_{m'g}}{(\omega_{gm}-\omega_3+i\Gamma_{gm})(\omega_{gm'}-\omega_2+i\Gamma_{gm'})}$$

$$+\frac{r^j_{gm}r^i_{mm'}r^k_{m'g}}{(\omega_{gm}+\omega_2+i\Gamma_{gm})(\omega_{gm'}+\omega_3+i\Gamma_{gm'})}$$

$$+\frac{r^i_{gm}r^k_{mm'}r^j_{m'g}}{(\omega_{gm}+\omega_2+i\Gamma_{gm})(\omega_{gm'}-\omega_1+i\Gamma_{gm'})}$$

$$+\frac{r^k_{gm}r^i_{mm'}r^j_{m'g}}{(\omega_{gm}-\omega_3+i\Gamma_{gm})(\omega_{gm'}+\omega_1+i\Gamma_{gm'})}). \tag{2.33}$$

Diagrams with corresponding expressions are a useful tool in analysis of various nonlinear processes. However, the expressions obtained are virtually useless for
calculations of susceptibilities of real materials. The reason is that it requires summation over an infinite number of states $m, m', \ldots$. These obviously include excited
states, which are difficult to obtain for any systems except very few simple atoms
and molecules. Actual application of this technique is known as Sum Over States

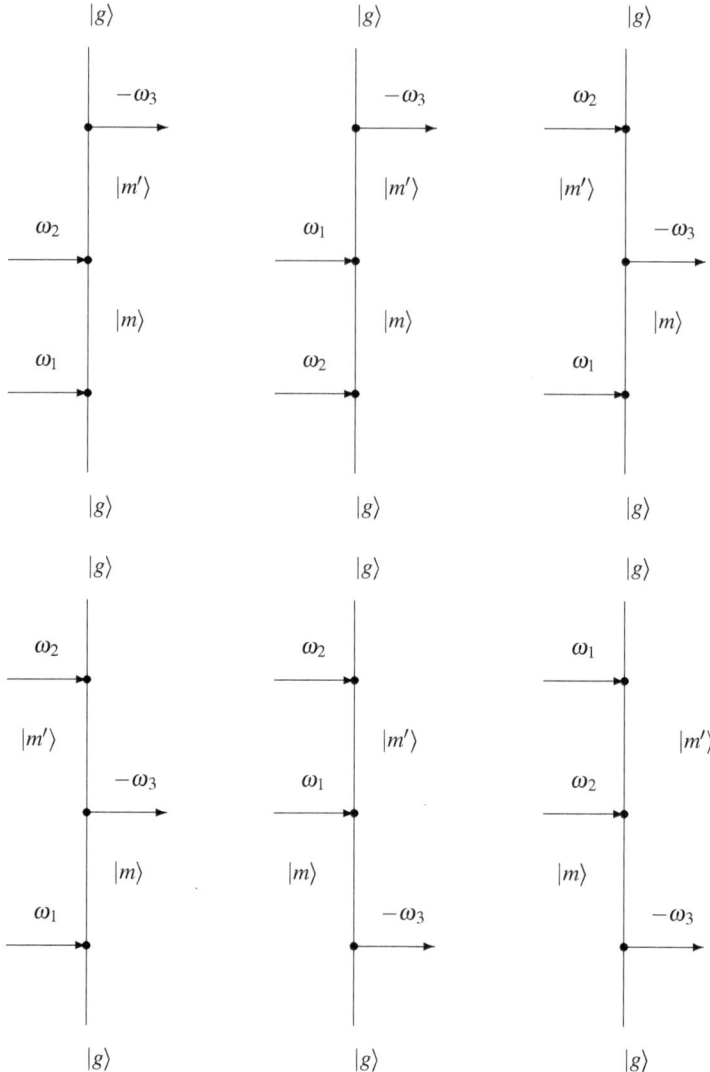

Fig. 2.2 Non-equivalent diagrams for second order process: $\chi^{(2)}(-\omega_3; \omega_1, \omega_2)$

approach, and involves additional approximations. A typical approximation is a trun-
cation of infinite summation to just a few states, sometimes as little as two or three.

References

1. L.D. Landau, E.M. Lifshitz, *Statistical Physics* (Butterworth-Heinemann, Oxford, 1975)
2. K.-E. Peiponen, J.J. Saarinen, Rep. Prog. Phys. **72**, 056401 (2009)
3. Y.R. Shen, *The Principles of Nonlinear Optics* (Wiley-Interscience, New York, 2001)
4. R.W. Boyd, *Nonlinear Optics* (Elsevier, New York, 2008)
5. A. Fetter, J. Walecka, *Quantum Theory of Many-Particle Systems* (McGraw-Hill, San Francisco, 1971)
6. N.B. Delone, V.P. Krainov, *Atoms in Strong Light Fields* (Springer, Berlin, 1985)
7. N.B. Delone, V.P. Krainov, *Fundamentals of Nonlinear Optics of Atomic Gases* (Wiley, New York, 1988)

Chapter 3
Density Functional Perturbation Theory

3.1 The Original Sternheimer Method

In 1954, R.M. Sternheimer used first order perturbation theory to calculate polarizabilities of closed-shell ions. The further development of his method evolved into what is now known as Density Functional Perturbation Theory (DFPT). Since DFPT is conceptually similar to his approach, it is instructive to start derivation of DFPT with the original work of Sternheimer [1].

Expanding Hamiltonian, energy and electron wavefunction in the first order perturbation theory one gets:

$$H(\mathbf{r}) = H_0(\mathbf{r}) + V_{ext}(\mathbf{r}), \tag{3.1}$$

$$E = E^{(0)} + E^{(1)}, \tag{3.2}$$

$$\psi(\mathbf{r}) = \psi^{(0)}(\mathbf{r}) + \psi^{(1)}(\mathbf{r}). \tag{3.3}$$

Here Hamiltonian is written as a sum of unperturbed Hamiltonian for a core electron $H_0(\mathbf{r})$, and external perturbation in dipole approximation $V_{ext}(\mathbf{r}) = -\frac{2e}{R^2} r \cos(\theta)$. R is a distance from a positive unit charge creating perturbing electric field at location of a core electron with radius vector $\mathbf{r}$. The ion core is at the origin, and θ is an angle between $\mathbf{R}$ and $\mathbf{r}$. First order energy correction $E^{(1)}$ is zero, because V_{ext} and $|\psi^{(0)}|^2$ are of different parity. Inserting (3.1)–(3.3) into stationary Schrödinger equation $H\psi = E\psi$,

$$H_0\psi^{(0)} + H_0\psi^{(1)} + V_{ext}\psi^{(0)} + o(2) = E^{(0)}\psi^{(0)} + E^{(0)}\psi^{(1)}$$

one gets to the first order:

$$(H_0 - E^{(0)})\psi^{(1)} = -V_{ext}\,\psi^{(0)}. \tag{3.4}$$

© The Author(s) 2014
V. Goncharov, *Non-Linear Optical Response in Atoms, Molecules and Clusters*,
SpringerBriefs in Electrical and Magnetic Properties of Atoms, Molecules, and Clusters,
DOI 10.1007/978-3-319-08320-9_3

Equation (3.4) is called Sternheimer equation. It is linear in respect to $\psi^{(1)}$, and is therefore highly amenable to efficient linear solvers. Solving (3.4) one gains access to the perturbed density:

$$\rho(\mathbf{r}) = |\psi^{(0)}(\mathbf{r}) + \psi^{(1)}(\mathbf{r})|^2 = |\psi^{(0)}(\mathbf{r})|^2 + 2\,\psi^{(0)}(\mathbf{r})\psi^{(1)}(\mathbf{r}) + o(2).$$

The first order density change $\delta\rho^{(1)} = \rho(\mathbf{r}) - \rho(\mathbf{r})^{(0)}$ is

$$\delta\rho^{(1)}(\mathbf{r}) = 2\,\psi^{(0)}(\mathbf{r})\psi^{(1)}(\mathbf{r}). \tag{3.5}$$

Now we can calculate the induced dipole moment of the core electron P_{ind}:

$$P_{ind} = -e\int \delta\rho^{(1)}(\mathbf{r})\, r\, dr^3 = \alpha E_{ext}, \tag{3.6}$$

and, because the perturbing electric field is known $E_{ext} = -\frac{e}{R^2}$, we can subsequently compute the static polarizability α:

$$\alpha = R^2 \int \delta\rho^{(1)}(\mathbf{r})\, r dr^3. \tag{3.7}$$

Sternheimer equation is a single electron equation. Yet, Mean Field Theories formulated in terms of single particle states interacting with "mean" potential such as Density Functional Theory (DFT), can take advantage of it. This is the reason why DFPT can retain the general structure of Sternheimer Equation (3.4), along with the steps in computing density change (3.5), polarization (3.6), and polarizability (3.7).

3.2 Modified Sternheimer Method

The original Sternheimer method does not include relaxation effects resulting from the change in the density of excited electron on the rest of electrons in the ionic core. In this sense it is an independent particle approximation. This leads to ~40 % discrepancy with experiment. In 1980, Stott and Zaremba [2], Zangwill and Soven [3] and Mahan [4] have developed methodology that may be called DFPT for electronic linear response in atoms. The method of Stott, Zaremba, Zangwill and Soven relied on Greens functions and was named by them the method of Self Consistent Field (SCF). Mahan on the other hand developed an extension to Sternheimer method and named it Modified Sternheimer (MS). In the next section we show that both methods are equivalent. Below, we follow Mahan's derivation.

Let's consider atomic electron with unperturbed LDA Hamiltonian

$$H_0(\mathbf{r}) = -\frac{\hbar^2 \nabla^2}{2m_e} + V(\mathbf{r}) \tag{3.8}$$

and Mean-Field potential in LDA form

$$V(\mathbf{r}) = -\frac{Z}{r} + \int dr^3 \frac{\rho(\mathbf{r}')}{|\mathbf{r} - \mathbf{r}'|} - V_{xc}[\rho(\mathbf{r})]. \tag{3.9}$$

Applying external perturbation $V_{ext}(\mathbf{r}) = r^l P_l(\cos(\theta))$ will lead to change in the wave function $\psi_i(\mathbf{r}) = \psi_i^{(0)}(\mathbf{r}) + \psi_i^{(1)}(\mathbf{r})$ that will manifest in density change:

$$\rho(\mathbf{r}) = \rho^{(0)}(\mathbf{r}) + \rho^{(1)}(\mathbf{r}) \tag{3.10}$$

$$\rho^{(1)}(\mathbf{r}) = 2\,\Re\left(\sum_i (\psi_i^{(0)}(\mathbf{r}))^* \psi_i^{(1)}(\mathbf{r})\right) \tag{3.11}$$

The density change in turn will induce the change in potential:

$$V(\mathbf{r}) \rightarrow V(\mathbf{r}) + V_{SCF}(\mathbf{r}), \tag{3.12}$$

where V_{SCF} is an induced self consistent field contribution:

$$V_{SCF}(\mathbf{r}) = V_{ext}(\mathbf{r}) + \int dr^3 \frac{\rho^{(1)}(\mathbf{r}')}{|\mathbf{r} - \mathbf{r}'|} + \rho^{(1)}(\mathbf{r})\left(\frac{\partial V_{xc}}{\partial \rho}\right)_{\rho^{(0)}}. \tag{3.13}$$

If the first order change in energy is negligible[1] $E_i \rightarrow E_i + o(2)$, then resulting equation is structurally identical to (3.4):

$$(H_0 - E_i)\psi_i^{(1)} = -V_{SCF}\,\psi_i^{(0)}. \tag{3.14}$$

The difference with original Sternheimer equation is that in (3.14) V_{SCF} replaces V_{ext} on the right hand side of (3.4). Solving (3.14) yields $\rho^{(1)}$, and allows one to calculate atomic polarizability, which in this case is

$$\alpha_l = 2a_B^{(2l+1)} \int dr^3\, r^l\, P_l(\cos(\theta))\, \rho^{(1)}(\mathbf{r}). \tag{3.15}$$

3.3 Greens Function Approach to DFPT

To analyze the meaning of Sternheimer equation we re-write (3.14) in operator form:

$$\hat{\Delta H_0}|\psi^{(1)}\rangle = |b\rangle. \tag{3.16}$$

---

[1] Otherwise, $\langle \psi^{(0)}|V_{SCF}|\psi^{(0)}\rangle$ is subtracted from V_{SCF}.

The operator $\hat{\Delta H}_0$ has the ground state of the system as its null-space:

$$\hat{\Delta H}_0|\psi^{(0)}\rangle = 0. \tag{3.17}$$

The vector $|b\rangle$ represents deviation of the ground state vector under the action of combined density and external field perturbation

$$|b\rangle = -\hat{V}_{SCF}|\psi^{(0)}\rangle. \tag{3.18}$$

Therefore, solution $|\psi^{(1)}\rangle$ to (3.16) (and of course to (3.14)) may be interpreted as a correction to the ground state due to the density response resulting from external perturbation. On another hand, if $\delta\rho(\mathbf{r})$ is a deviation from ground state density arising from $\hat{V}_{SCF}$, then it could be written within the Linear Response Theory in terms of the ground state density–density response function $\Xi_{(0)}(\mathbf{r},\mathbf{r}')$

$$\delta\rho(\mathbf{r}) = \int dr'^3 \Xi_{(0)}(\mathbf{r},\mathbf{r}')V_{SCF}(\mathbf{r}'). \tag{3.19}$$

The Linear density–density response may be expressed through the sum over occupied Kohn-Sham orbitals and Green's function.

$$\Xi_{(0)}(\mathbf{r},\mathbf{r}';\omega) = \sum_i^{occ.}\{\psi_i^*(\mathbf{r})\psi_i(\mathbf{r})G^{(+)}(\mathbf{r},\mathbf{r}';\varepsilon_i+\hbar\omega)$$
$$+ \psi_i(\mathbf{r})\psi_i^*(\mathbf{r})G^{(-)}(\mathbf{r}',\mathbf{r};\varepsilon_i-\hbar\omega)\} \tag{3.20}$$

$$G^{(\pm)}(\mathbf{r},\mathbf{r}';E) = \sum_j^\infty \frac{\psi_j(\mathbf{r})\psi_j^*(\mathbf{r})}{E-\varepsilon_j\pm i\eta} \tag{3.21}$$

This approach suffers from a complication with the summation of over all states in (3.21). The density–density response function could be written exclusively in terms of single particle Greens function:

$$\Xi_{(0)}(\mathbf{r},\mathbf{r}') = -\frac{2}{\pi}\Im\left(\int_{-\infty}^{\mu} d\omega G^+(\mathbf{r},\mathbf{r}',\omega)G^+(\mathbf{r}',\mathbf{r},\omega)\right). \tag{3.22}$$

To avoid complications of the sum-over states representation, the Greens function could be obtained directly from

$$\left(-\frac{\hbar^2\nabla^2}{2m_e}+V(\mathbf{r})-\omega\right)G^+(\mathbf{r},\mathbf{r}',\omega) = -\delta(\mathbf{r}-\mathbf{r}'). \tag{3.23}$$

It this approach, the perturbation could be easily generalized to frequency dependent case. Then the density response becomes:

$$\delta\rho(\mathbf{r}, \omega) = \int dr'^3 \, \Xi_{(0)}(\mathbf{r}, \mathbf{r}', \omega) V_{SCF}(\mathbf{r}', \omega). \tag{3.24}$$

Knowing solution to (3.23), one computes density response function from (3.22), then density response from (3.24), and finally polarizability. This is exactly the SCF method of Stott, Zaremba, Zangwill and Soven. From derivation it is clear that SCF is equivalent to Modified Sternheimer equation.

3.4 Dyson Equation of DFPT

The density–density response function appearing in (3.24) represents non-interacting ground state within DFT. It is also called Kohn-Sham density–density response function. If we knew the exact, interacting density–density response function Ξ^2, then instead "of self-consistent" potential V_{SCF} we would have V_{ext} as perturbing potential:

$$\rho^{(1)}(\mathbf{r}, \omega) = \int dr'^3 \, \Xi(\mathbf{r}, \mathbf{r}', \omega) V_{ext}(\mathbf{r}', \omega). \tag{3.25}$$

Expanding V_{SCF} in (3.24) we get:

$$\rho^{(1)}(\mathbf{r}, \omega) = \int dr'^3 \, \Xi_{(0)}(\mathbf{r}, \mathbf{r}', \omega) V_{ext}(\mathbf{r}', \omega) + \int dr'^3 \, \Xi_{(0)}(\mathbf{r}, \mathbf{r}', \omega)$$
$$\times \int dr'^3 \left(\frac{e^2}{|\mathbf{r} - \mathbf{r}'|} + f_{xc}(\mathbf{r}, \mathbf{r}'; \omega) \right) \rho^{(1)}(\mathbf{r}', \omega)(\mathbf{r}', \omega) \tag{3.26}$$

where f_{xc} is the first order kernel:

$$f_{xc}(\mathbf{r}, \mathbf{r}'; \omega) = \left(\frac{\partial V_{xc}}{\partial \rho} \right)_{\rho^{(0)}} \delta(\mathbf{r} - \mathbf{r}'). \tag{3.27}$$

Since $\rho^{(1)}$ appears on both sides of the integral equation, computing it is not as straightforward as it may first appear. Typically, it is solved by a method of self-consistent iteration, similarly to Kohn-Sham ground state problem. Comparing (3.25) and (3.26) one may arrive at the equation for the exact, interacting density–density response function:

---

[2] Also called reducible polarization.

$$\Xi(\mathbf{r}, \mathbf{r}', \omega) = \Xi_{(0)}(\mathbf{r}, \mathbf{r}', \omega) + \int dr''^3 \int dr'^3 \; \Xi_{(0)}(\mathbf{r}, \mathbf{r}'')$$

$$\times \left(\frac{e^2}{|\mathbf{r}' - \mathbf{r}''|} + f_{xc}(\mathbf{r}', \mathbf{r}'', \omega) \right) \Xi(\mathbf{r}'', \mathbf{r}', \omega). \tag{3.28}$$

This equation is called Dyson Equation, and it provides an additional, direct approach to computing linear response within DFPT. This equation shows in a particularly transparent way that the fidelity of the response calculations within DFPT is critically dependent on the "correctness" of the exchange-correlation functional.

3.5 Dynamic Sternheimer Equation

Explicitly time dependent external potential can also be treated within Sternheimer framework. The extension could be achieved by an ansatz for the first order corrections to the wavefunctions $\psi_n^{(1)}$:

$$\psi_n^{(1)}(\mathbf{r}) \rightarrow e^{\pm i \omega t} \psi_n^{(1)}(\mathbf{r}, \pm\omega), \tag{3.29}$$

and simultaneous replacement of the static potential with Fourier transform of the time dependent potential:

$$V_{ext}(\mathbf{r}) \rightarrow V_{ext}(\mathbf{r}, \pm\omega). \tag{3.30}$$

Inserting (3.29) and (3.30) into $i\hbar \frac{\partial \psi_n}{\partial t} = H \psi_n$ one obtains:

$$\left(H[\rho] - \varepsilon_n^{(0)} \pm \hbar\omega + i\eta \right) \psi_n^{(1)}(\mathbf{r}, \pm\omega) = -V_{SCF}(\mathbf{r}, \pm\omega) \psi_n^{(0)}(\mathbf{r}). \tag{3.31}$$

In practice (3.31) is first solved in the space orthogonal to the ground state orbitals:

$$\left(H[\rho] - \varepsilon_n^{(0)} \pm \hbar\omega + i\eta \right) \psi_n^{(1)}(\mathbf{r}, \pm\omega) = -V_{SCF}(\mathbf{r}, \pm\omega) \psi_n^{(0)}(\mathbf{r})$$

$$- \sum_j^{occ} \psi_j^{(0)}(\mathbf{r}) \langle \psi_j^{(0)} | V_{SCF} | \psi_n^{(0)} \rangle, \tag{3.32}$$

and then the ground state components are added:

$$\psi_n^{(1)}(\mathbf{r}, \pm\omega) = \psi_n^{(1)}(\mathbf{r}, \pm\omega) + \sum_j^{occ} \frac{\psi_j^{(0)}(\mathbf{r}) \langle \psi_j^{(0)} | V_{SCF} | \psi_n^{(0)} \rangle}{\varepsilon_n^{(0)} - \varepsilon_j^{(0)} \pm \hbar\omega} \tag{3.33}$$

Symbolically this could be written with the use of projection operator $\mathbf{P}_{\perp\{\psi_n^{(0)}\}}$:

$$\left(H[\rho] - \varepsilon_n^{(0)} \pm \hbar\omega + i\eta\right) \psi_n^{(1)}(\mathbf{r}, \pm\omega) = -\mathbf{P}_{\perp\{\psi_n^{(0)}\}} V_{SCF}(\mathbf{r}, \pm\omega)\, \psi_n^{(0)}(\mathbf{r}). \quad (3.34)$$

3.6 Sternheimer Method for Nonlinear Response

In this section we extend formalism to an arbitrary finite system that includes molecules, nanoparticles etc.. The relevant single particle Hamiltonian is:

$$H(\mathbf{r}) = -\frac{\hbar^2 \nabla^2}{2m_e} + e^2 \int dr^3 \frac{\rho(\mathbf{r}')}{|\mathbf{r} - \mathbf{r}'|} + V_{xc}[\rho(\mathbf{r}), t]) + V_{ext}(\mathbf{r}, t) + V_{ion}. \quad (3.35)$$

The ions are presumed immobile and their contribution is represented by a potential V_{ion}. The external potential is taken in a dipole approximation as $V_{ext}(\mathbf{r}, \omega) = -\sum_{i=1}^{3} eE_i(\omega)r_i$, i.e. atomic scale local field effects are not considered within this formalism. The time dependent density and effective perturbation V_{SCF} are formally written as a series up to the highest order of interest[3]:

$$\rho(\mathbf{r}, t) = \rho(\mathbf{r})^{(0)} + \rho^{(1)}(\mathbf{r}, t) + \rho^{(2)}(\mathbf{r}, t) + \cdots \quad (3.36)$$

$$H[\rho](\mathbf{r}, t) = H[\rho^{(0)}](\mathbf{r}) + V^{(1)}(\mathbf{r}, t) + V^{(2)}(\mathbf{r}, t) + \cdots$$

The zero order density is static. The first orders correspond to LR. The higher orders representing NLR are in practice truncated at the third order. The formalism is developed in Frequency Domain:

$$\rho^{(k)}(\mathbf{r}, t) = \frac{1}{2\pi} \int\limits_{-\infty}^{+\infty} \rho^{(k)}(\mathbf{r}, \omega)\, e^{-i\omega t}\, d\omega$$

$$V^{(k)}(\mathbf{r}, t) = \frac{1}{2\pi} \int\limits_{-\infty}^{+\infty} V^{(k)}(\mathbf{r}, \omega)\, e^{-i\omega t}\, d\omega$$

$$V_{ext}(\mathbf{r}, t) = \frac{1}{2\pi} \int\limits_{-\infty}^{+\infty} V_{ext}(\mathbf{r}, \omega)\, e^{-i\omega t}\, d\omega$$

---

[3] In this case the second order.

The frequency dependent density variations $\rho^{(n)}$ are decomposed into sums of transition densities $\rho^{(n)}_{j...n}(\mathbf{r}, -\omega_\sigma; \omega_1 ... \omega_n)$ defined via:

$$\rho^{(n)}(\mathbf{r}, \omega_\sigma) = \frac{N_{perm}}{(2\pi)^{n-l-m} n!} \sum_{j...n} \int_{-\infty}^{+\infty} \rho^{(n)}_{j...n}(\mathbf{r}, -\omega_\sigma; \omega_1 ... \omega_n)$$

$$\times \; \delta(\omega_1 + \cdots + \omega_n - \omega_\sigma) \, d\omega_1 ... d\omega_n, \tag{3.37}$$

where n is order of process, N_{perm}—number of frequencies ω_k permutations, m-number of frequencies equal to zero, and l is zero if $\omega_\sigma = 0$, otherwise it is 1. The delta function insures conservation of energy. The transition densities are directly related to optical response functions linear polarizability and hyperpolarizabilities:

$$\chi^{(n)}_{ij...n}(-\omega_\sigma; \omega_1 ... \omega_n) = -e \int d\mathbf{r} \; r_i \; \frac{\rho^{(n)}_{j...n}(\mathbf{r}, -\omega_\sigma; \omega_1 ... \omega_n)}{E_j(\omega_1) ... E_n(\omega_n)}. \tag{3.38}$$

First order transition density is similar to (3.24):

$$\rho^{(1)}_j(\mathbf{r}, \omega) = \int d\mathbf{r}' \; \Xi^{(1)}(\mathbf{r}, \mathbf{r}'; \omega) \; V^{(1)}_j(\mathbf{r}', \omega). \tag{3.39}$$

Similarly to (3.37), we also define transition perturbation $V^{(n)}_{j...n}(\mathbf{r}, -\omega_\sigma; \omega_1 ... \omega_n)$. Transition perturbations are generally obtained by differentiating Hamiltonian, except for the linear order. For the first order from $V^{(1)}_j = \frac{\partial H}{\partial \rho} \rho^{(1)}_j + (V_{ext})_j$ we obtain

$$V^{(1)}_j(\mathbf{r}, \omega) = \int dr'^3 \left(\frac{e^2}{|\mathbf{r} - \mathbf{r}'|} + f_{xc}(\mathbf{r}, \mathbf{r}'; \omega) \right) \rho^{(1)}_j(\mathbf{r}', \omega) - eE_j(\omega)r_j, \tag{3.40}$$

Inserting (3.40) into (3.39) we get expression similar to (3.26):

$$\rho^{(1)}_j(\mathbf{r}, \omega) = \int d\mathbf{r}' \; \Xi^{(1)}(\mathbf{r}, \mathbf{r}'; \omega) \left[\int dr'^3 \left(\frac{e^2}{|\mathbf{r} - \mathbf{r}'|} + f_{xc}(\mathbf{r}, \mathbf{r}'; \omega) \right) \rho^{(1)}_j(\mathbf{r}', \omega) \right.$$
$$\left. - eE_j(\omega)r_j \right] \tag{3.41}$$

This equation has to be solved self consistently. Physically, solving Eq. (3.41) is equivalent to solving (3.34). The computational approach is described in the next section.

Similarly, for the second order perturbation $V^{(2)}_{jk} = \frac{\partial^2 H}{\partial \rho^2} \rho^{(1)}_j \rho^{(1)}_j + \frac{\partial H}{\partial \rho} \rho^{(2)}_{jk}$ one gets

$$V_{jk}^{(2)}(\mathbf{r}, \omega; \omega_1, \omega_2) = \int dr'^3 \left(\frac{e^2}{|\mathbf{r} - \mathbf{r}'|} + f_{xc}(\mathbf{r}, \mathbf{r}'; \omega) \right) \rho_{jk}^{(2)}(\mathbf{r}', -\omega; \omega_1, \omega_2)$$

$$+ \int dr'^3 \, dr''^3 g_{xc}(\mathbf{r}, \mathbf{r}', \mathbf{r}''; \omega_1, \omega_2) \rho_j^{(1)}(\mathbf{r}', \omega_1) \rho_k^{(1)}(\mathbf{r}'', \omega_2).$$

$$(3.42)$$

where g_{xc} is second order kernel:

$$g_{xc}(\mathbf{r}, \mathbf{r}', \mathbf{r}''; \omega_1, \omega_2) = \left(\frac{\partial^2 V_{xc}}{\partial \rho^2} \right)_{\rho^{(0)}} \delta(\mathbf{r} - \mathbf{r}')\delta(\mathbf{r} - \mathbf{r}''). \qquad (3.43)$$

The second order transition density consists of two second order terms. One term represents the first order response from the second order transition perturbation, and the other term is the second order response from the term which is quadratic in the first order transition perturbation:

$$\rho_{jk}^{(2)}(\mathbf{r}, -\omega; \omega_1, \omega_2) = \int dr'^3 \, \Xi^{(1)}(\mathbf{r}, \mathbf{r}'; \omega) V_{jk}^{(2)}(\mathbf{r}', -\omega; \omega_1, \omega_2)$$

$$+ \int dr'^3 \, dr''^3 \, \Xi^{(2)}(\mathbf{r}, \mathbf{r}', \mathbf{r}'', -\omega; \omega_1, \omega_2)$$

$$\times V_j^{(1)}(\mathbf{r}', \omega_1) V_k^{(1)}(\mathbf{r}'', \omega_2) \qquad (3.44)$$

Combining these one gets:

$$\rho_{jk}^{(2)}(\mathbf{r}, -\omega; \omega_1, \omega_2) = \int dr'^3 \, \Xi^{(1)}(\mathbf{r}, \mathbf{r}'; \omega) [\int dr'''^3 \left(\frac{e^2}{|\mathbf{r}' - \mathbf{r}'''|} + f_{xc}(\mathbf{r}', \mathbf{r}'''; \omega) \right)$$

$$\times \rho_{jk}^{(2)}(\mathbf{r}''', -\omega; \omega_1, \omega_2) + \int dr'''^3 \, dr''^3 g_{xc}(\mathbf{r}', \mathbf{r}''', \mathbf{r}''; \omega_1, \omega_2)$$

$$\times \rho_j^{(1)}(\mathbf{r}''', \omega_1) \rho_k^{(1)}(\mathbf{r}'', \omega_2)] + \int dr'^3 \, dr''^3 \, \Xi^{(2)}$$

$$\times (\mathbf{r}, \mathbf{r}', \mathbf{r}'', -\omega; \omega_1, \omega_2) \times V_j^{(1)}(\mathbf{r}', \omega_1) V_k^{(1)}(\mathbf{r}'', \omega_2) \qquad (3.45)$$

Left hand side of (3.45) and the first term on right hand side depend on second order transition density $\rho_{jk}^{(2)}$. However, other terms on the right hand side depend only on the first order transition density $\rho_{jk}^{(1)}$, and therefore can be computed independently during the preceding step from (3.41). Moreover, the last term on right hand side contains ground state second order density–density response function $\Xi^{(2)}$ and is computed non self-consistently. It is clear that the method must be recursive, because higher orders are dependent on the calculation of lower ones. Further, the equations for higher order transition density maybe written as:

$$\rho_{j\ldots n}^{(n)}(\mathbf{r}, -\omega_\sigma; \omega_1 \ldots \omega_n) = \int dr'^3 \; \Xi^{(1)}(\mathbf{r}, \mathbf{r}'; \omega_\sigma) \int dr''^3 \frac{e^2}{|\mathbf{r} - \mathbf{r}'|} + f_{xc}(\mathbf{r}', \mathbf{r}''; \omega_\sigma)$$

$$\times \rho_{j\ldots n}^{(n)}(\mathbf{r}'', -\omega_\sigma; \omega_1 \ldots \omega_n)$$

$$+ \mathscr{F}[\rho_{j\ldots n-1}^{(n-1)}(\mathbf{r}, -\omega_\sigma; \omega_1 \ldots \omega_{n-1})]. \tag{3.46}$$

Here $\mathscr{F}[\rho_{j\ldots n-1}^{(n-1)}$ is the sum of terms that depends on $(n-1)$ transition density. The structure of equation for the nth density response (3.46) is essentially the same as for the first order, except the $\mathscr{F}$ term, which can be computed immediately before self consistent iteration. This means we can use essentially the same Sternheimer procedure for solving nth response as in the linear case.

3.7 Algorithm for Solving DFPT Equations

Sternheimer equation allows us to circumvent explicit calculation of the density–density response functions in equations of type (3.46). The algorithm relies on two basic iterative routines. We will first give an algorithmic description of these iterators, and then show how to use them as building blocks in constructing a sequence for calculating polarizability and subsequently hyperpolarizabilities. Both routines are aimed at solving a linear system of a particular structure. One procedure is required for solution of linear equation that has a structure of Sternheimer Equation (3.34). We call it Sternheimer equation iteration (SEI). It is solved by one of the many non stationary iterative solvers that include Conjugate Gradient, BiConjugate Gradient and Conjugate Gradient Squared methods. The second building block is a self consistent iteration (SCI) used for the solution of linear equation that has the structure of (3.39). SCI relies on a non-symmetric linear solver such as Generalized Conjugate Residuals method. The computation of density–density response function $\Xi^{(1)}$ that appears in (3.39) is avoided by solving corresponding Sternhemer Equation at each iteration.

3.7.1 Sternheimer Equation Iteration

The SEI procedure involves the following steps:

(1) For the given input energy E and transition perturbation $V_{x_1\ldots x_{k-1}}^{(k-1)}$ use CG algorithm to solve Sternheimer equation for $\psi_p^{(+)}$:

$$\left(H[\rho] - \varepsilon_p^{(0)} - E\right) \psi_p^{(+)} = -\mathbf{P}_{\perp\{\psi_n^{(0)}\}} V_{x_1\ldots x_{k-1}}^{(k-1)} \psi_p^{(0)}$$

(2) Repeat step 1 for $\psi_p^{(-)}$:

$$\left(H[\rho] - \varepsilon_p^{(0)} + E\right) \psi_p^{(-)} = -\mathbf{P}_{\perp\{\psi_n^{(0)}\}} V_{x_1...x_{k-1}}^{(k-1)} \psi_p^{(0)}$$

(3) Compute kth order transition density change:

$$\delta\rho_{x_1...x_{k-1}}^{(k)} = \sum_p (\psi_p^{(0)} \psi_p^{(+)} + \psi_p^{(0)} \psi_p^{(-)})$$

The SEI procedure is very similar to particle-hole excitation calculations in Random Phase Approximation. It is used both in SCI procedure and in the construction of higher order density–density response function $\Xi^{(k)}$.

3.7.2 Self Consistent Iteration

For an initial transition density $\rho_{x_1...x_k}^{(init)}$ and energy E, SCI procedure computes kth order transition density $\rho_{x_1...x_k}^{(k)}$. It involves the following steps:

Initialize $p_0 = r_0 = \rho_{x_1...x_k}^{(init)}$; $Ap_0 = \rho^{(init)} - \rho^{(k)}$; $X_0 = 0$.

For $n = 0, \ldots, N_{max}$

(1)
$$\alpha_n = \frac{r_n \, A \, p_n}{\langle A \, p_n \, A \, p_n \rangle}$$

(2)
$$X_n = X_n + \alpha_n \, p_n$$

(3)
$$r_n = r_n - \alpha_n \, A \, p_n$$

(4)
$$v_n = \int \left(\frac{1}{|\mathbf{r} - \mathbf{r}'|} + f_{xc}(\mathbf{r}', \mathbf{r})\right) r_n \, d\,r'^3$$

(5) Invoke SEI to solve

$$\left(H[\rho] - \varepsilon_p^{(0)} \pm E\right) \psi_p^{(\pm)} = -v_n \, \psi_p^{(0)}$$

to obtain $\rho^{(k)}$ for given energy E and potential v_n, then

$$\delta\rho_n'' = r_n - \rho_{x_1...x_k}^{(k)}$$

(6)

$$\beta_n = \frac{\delta \rho_n'' \, A \, p_n}{\langle A \, p_n \, A \, p_n \rangle}$$

(7)

$$p_{n+1} = r_n + \beta_n \, p_n$$

(8)

$$A \, p_{n+1} = \delta \rho_n'' + A \, p_n$$

(9) If X_n converged, then let $\rho_{j...k}^{(k)} = X_n$, and exit cycle.

3.7.3 Linear Polarizability

Polarizability at the frequency of interest ω_0 is calculated as follows:
First, use SEI to solve

$$\left(H[\rho] - \varepsilon_p^{(0)} \pm \hbar \omega_0 \right) \psi_p^{(\pm)} = -r_\mu \, \psi_p^{(0)}$$

for given energy $\hbar \omega_0$ and potential $r_\mu \, E_\mu$. This gives zero-order, non-self consistent $\rho_\mu^{(0)}$ that is used as a starting transition density in the SCI. Next, SCI yields the first order transition density $\rho_\mu^{(1)}$ which is used to calculate polarizability

$$\alpha_{\mu\mu}(\omega_0) = -e \int dr^3 \, r_\mu \, \frac{\rho_\mu^{(1)}}{E_\mu}.$$

These steps are repeated for each Cartesian component: $\mu = x, y, y$. Finally, average polarizability is obtained: $\alpha_{avg}(\omega_0) = \frac{1}{3}(\alpha_{xx} + \alpha_{yy} + \alpha_{zz})$.

3.7.4 First Hyperpolarizability

Here we show how to calculate hyperpolarizability $\beta(\pm\omega_1 \pm\omega_2)$. To start calculations one needs first order transition perturbation $V_\mu^{(1)}$ which is computed from transition density $\rho_\mu^{(1)}$:

$$V_\mu^{(1)}(\mathbf{r}, \omega_i) = \int dr'^3 \left(\frac{e^2}{|\mathbf{r} - \mathbf{r}'|} + f_{xc}(\mathbf{r}, \mathbf{r}'; \omega) \right) \rho_\mu^{(1)}(\mathbf{r}', \omega) - e E_\mu(\omega) r_\mu.$$

Using $V^{(1)}$ construct second order density–density response function $\Xi^{(2)}$:

(1) Use SEI to solve

$$\left(H[\rho] - \varepsilon_p^{(0)} - \hbar\omega_1\right) \psi_{p\mu}^{(1+)} = -V_\mu^{(1)}(\mathbf{r}, \omega_1)\, \psi_p^{(0)}$$

for given energy $\hbar\omega_1$ and potential $V_\mu^{(1)}(\mathbf{r}, \omega_1)$

(2) Similarly solve

$$\left(H[\rho] - \varepsilon_p^{(0)} + \hbar\omega_2\right) \psi_{p\nu}^{(1-)} = -V_\nu^{(1)}(\mathbf{r}, \omega_2)\, \psi_p^{(0)}$$

(3) Similarly solve

$$\left(H[\rho] - \varepsilon_p^{(0)} + \hbar\omega_1\right) \psi_{p\mu}^{(1-)} = -V_\mu^{(1)}(\mathbf{r}, \omega_1)\, \psi_p^{(0)}$$

(4) Solve

$$\left(H[\rho] - \varepsilon_p^{(0)} + \hbar\omega_1 + \hbar\omega_2\right) \psi_{p\mu\nu}^{(2-)} = -V_\mu^{(1)}(\mathbf{r}, \omega_1)\, \psi_{p\nu}^{(1-)}$$

(5) Solve

$$\left(H[\rho] - \varepsilon_p^{(0)} - \hbar\omega_1 - \hbar\omega_2\right) \psi_{p\nu\mu}^{(2+)} = -V_\nu^{(1)}(\mathbf{r}, \omega_1)\, \psi_{p\mu}^{(1+)}$$

(6) Calculate contribution to the second order transition density $\Delta_2\rho_{\mu\nu}^{(2)}$:

$$\Delta_2\rho_{\mu\nu}^{(2)}(\mathbf{r}; \pm\omega_1 \pm \omega_2) = \sum_p (\psi_p^{(0)} \psi_{p\nu\mu}^{(2+)}(\omega_1 + \omega_2)$$

$$+ \psi_p^{(0)} \psi_{p\mu\nu}^{(2-)}(-\omega_1 - \omega_2) + \psi_{p\mu}^{(1+)}(\omega_1)\psi_{p\nu}^{(1-)}(\omega_2))$$

$\Delta_2\rho_{\mu\nu}^{(2)}$ is the second term on the left hand side of (3.45). Next, one calculates contribution to the second order transition density contribution $\Delta_1\rho_{\mu\nu}^{(2)}$. First, one solves

$$\left(H[\rho] - \varepsilon_p^{(0)} \pm \hbar\omega_1 \pm \hbar\omega_2\right) \psi_{p\nu\mu}^{(2\pm)} = -(g_{xc}(\omega_1, \omega_2)\rho_\mu^{(1)}(\omega_1)\rho_\nu^{(1)}(\omega_2))\, \psi_p^{(0)}.$$

Then one computes

$$\Delta_1\rho_{\mu\nu}^{(2)}(\mathbf{r}; \pm\omega_1 \pm \omega_2) = \sum_p \psi_p^{(0)} \psi_{p\nu\mu}^{(2+)}(\pm\omega_1 \pm \omega_2).$$

$\Delta_1\rho_{\mu\nu}^{(2)}$ is the first term on the right hand side of (3.45). The quantity $\Delta_1\rho_{\mu\nu}^{(2)} + \Delta_2\rho_{\mu\nu}^{(2)}$ is an initial value of the second order transition density that is used in SCI to compute $\rho_{\mu\nu}^{(2)}$. Once $\rho_{\mu\nu}^{(2)}$ is known, the first hyperpolarizability is computed using

Table 3.1 Polarizabilities and hyperpolarizabilities of small molecules calculated using modified Sternheimer method

Molecule	$\alpha(0)$	$\alpha(\omega)$	$\beta(0)$	SHG	EFISH	EOKE
H_2	5.85	5.90	0	0	1145	1080
CO	13.55	13.65	30.63	32.20	2427	2245
H_2O	10.43	10.52	-28.99	-31.38	3574	3225
C_2H_4	28.45	28.74	0	0	10466	9280

$\alpha(0)$, $\beta(0)$ are static polarizability and first hyperpolarizability. The frequency $\omega = 1.175\,\text{eV}$. All other data is in atomic units

$$\beta_{\eta\mu\nu}(\pm\omega_1 \pm \omega_2) = -e \int dr^3 \, r_\eta \, \frac{\rho^{(2)}_{\mu\nu}(\pm\omega_1 \pm \omega_2)}{E_\mu E_\nu}.$$

3.8 Illustration of the Modified Sternheimer Method

First, the ground state was calculated at the LDA level using Conjugate Gradients minimization of the total electronic energy. Approximately 200 iterations were used to insure that the single particle energies converged better then 0.01 eV. Perdew–Zunger exchange correlation functional has been used. Troulier–Martins pseudopotentials were used to represent the combined effect of the closed shell electrons and ionic cores. All calculations were done on cubic uniform grid of size 7 and 0.25 Å grid step. It has been shown in work of Iwata et al. that these parameters are sufficient for convergence [1].

Next, the Sternheimer algorithm described in the previous section has been used recursively. At first level the static and the dynamic polarizabilities at ω were calculated. At the second level the first static and the dynamic SHG hyperpolarizabilities were calculated. Finally, second hyperpolarizabilities corresponding to EFISH $\chi^{(3)}(-2\omega; 0, \omega, \omega)$, and to EOKE $\chi^{(3)}(-\omega; 0, 0, \omega)$ were calculated (Table 3.1).

References

1. R.M. Sternheimer, Phys. Rev. **96**, 951 (1954)
2. M.J. Stott, E. Zaremba, Phys. Rev. A **21**, 12 (1980)
3. A. Zangwill, P. Soven, Phys. Rev. A **21**, 1561 (1980)
4. G.D. Mahan, Phys. Rev. A **22**, 1780 (1980)
5. J.-I. Iwata, K. Yabana, G.F. Bertsch, J. Chem. Phys. **115**, 8773 (2001)

Chapter 4
Real Time Method

4.1 Adiabatic Local Density Approximation

Since the optical susceptibilities describe the light-matter interactions, their realistic evaluation at certain point requires solution of Schrödinger equation for a multi-electron system. All important results of this work were obtained by using time evolution of wavefunction under action of time dependent Hamiltonian in the Adiabatic Local Density Approximation.[1] In this approximation, the electron-electron interactions are described by effective density dependent potential $V_{eff}(\mathbf{r}, \rho(\mathbf{r}, t))$. The electronic density ρ is calculated from single particle orbitals ϕ:

$$\rho(\mathbf{r}, t) = \sum_{i=1}^{N} \phi_i^*(\mathbf{r}, t)\phi_i(\mathbf{r}, t), \tag{4.1}$$

where N is a number of electrons. ALDA involves three important approximations. First, it represents multi electron wave function as a determinant composed of single particle orbitals. This step reduces electron-electron interactions to two parts. One of them is a local, density dependent Hartree potential $V_H(\mathbf{r}, t)$:

$$V_H(\mathbf{r}, \rho(\mathbf{r}, t)) = \int d^3 r' \frac{\rho(\mathbf{r}', t)}{|\mathbf{r} - \mathbf{r}'|}. \tag{4.2}$$

Second, the exchange interaction is approximated by another local, density dependent potential $V_x(\mathbf{r}, \rho(\mathbf{r}, t))$:

$$V_x(\mathbf{r}, \rho(\mathbf{r}, t)) = -\frac{4}{3}C_x \rho(\mathbf{r}, t)^{\frac{1}{3}}, \tag{4.3}$$

---

[1] A number of textbooks has been published on this topic [1–3].

© The Author(s) 2014
V. Goncharov, *Non-Linear Optical Response in Atoms, Molecules and Clusters*,
SpringerBriefs in Electrical and Magnetic Properties of Atoms, Molecules, and Clusters,
DOI 10.1007/978-3-319-08320-9_4

where C_x is a constant. Then, a gamut of all correlations beyond spin exchange are represented by yet another local, density dependent correlation potential $V_c(\mathbf{r}, \rho(\mathbf{r}, t))$. This step modifies the first step in a way that the wavefunction now captures interactions that are not accessible to a single determinant states even with the exact exchange. As the result, electron-electron interactions in ALDA are represented by a sum of local density dependent potentials: Hartree and Exchange-Correlation.

$$V_{eff}(\mathbf{r}, t) = V_H(\mathbf{r}, \rho(\mathbf{r}, t)) + V_{xc}(\mathbf{r}, \rho(\mathbf{r}, t)). \qquad (4.4)$$

Above, the exchange and correlation potentials are written as a single potential $V_{xc} = V_x + V_c$. The theoretical foundations of ALDA are rooted in Time Dependent Density Functional Theory [4]. The limits of applicability of TDDFT is a hotly contested topic. Two common variants of ALDA exchange-correlation potentials are based on works of Perdew and Zunger [5] and Vosko et al. [6]. The potentials are named after them and are referred below as PZ and VWN. Next section introduces solution of time-dependent Schrödinger equation.

4.2 Real-Time Evolution

In order to obtain explicitly time dependent wavefunction on the level of ALDA, one needs to solve time dependent single particle Schödinger equation for a time dependent Hamiltonian with external potential

$$\hat{H}(t) = -\frac{\hbar^2}{2\,m_e}\Delta + V_{eff}(\mathbf{r}, s, t) + V_{ext}(\mathbf{r}, t) \qquad (4.5)$$

Ground state is obtained from solution of Kohn Sham equations:

$$\hat{H}(t_0)|\phi_i(t_0)\rangle = E_i|\phi_i(t_0)\rangle \qquad (4.6)$$

Starting from ground state, orbitals are evolved in time under the action of evolution operator $\hat{U}$

$$|\phi_i(t)\rangle = \hat{U}(t, t_0)|\phi_i(t_0)\rangle, \qquad (4.7)$$

Evolution operator is defined as time ordered exponential operator:

$$\hat{U}(t, t_0) = \hat{T}exp\left(\frac{-i}{\hbar} \int_{t_0}^{t} \hat{H}(t')dt' \right)$$

$$= \sum_{n=0}^{\infty} \frac{(-i)^n}{\hbar^n} \frac{1}{n!} \int_{t_0}^{t} dt_1 \ldots \int_{t_0}^{t} dt_n \hat{T}(\hat{H}(t_1) \ldots \hat{H}(t_n)) \qquad (4.8)$$

To utilize multiplicative property $\hat{U}(t_2, t_0) = \hat{U}(t_2, t_1)\hat{U}(t_1, t_0)$ finite time interval is broken into N_{time} small segments δt. Evolution operator becomes a product of operators applied at each $t_i + \delta t$:

$$\hat{U}(t, t_1) = \prod_{i=1}^{N_{time}} \hat{U}(t_i + \delta t, t_i) \tag{4.9}$$

Each δt corresponds to a evolution iteration at which Hamiltonian is taken constant with a value at that instant. Integral turns into a simple product:

$$\int_{t_i}^{t_i+\delta t} \hat{H}(t_i)dt \;\rightarrow\; \hat{H}(t_i)\delta t$$

In addition, the infinite sum is truncated at some highest order N_{taylor} turning the exponential into a Taylor polynomial. Therefore, the evolution operator for "physically infinitesimal" segment δt becomes:

$$\hat{U}(t_i + \delta t, t_i) = \sum_{n=0}^{N_{taylor}} \frac{1}{n!} \left(\frac{-i\,\hat{H}(t_i)\,\delta t}{\hbar} \right)^n \tag{4.10}$$

4.3 Real-Space Implementation

In this section we give details on numerical implementation of the real-time TDDFT in real-space. In real-space calculations [7], functions and operators that depend on spatial coordinates are represented on a three dimensional lattice. Differential operators that act on spatial variables become finite-difference operators. Action of the Hamiltonian on an orbital ϕ_q is represented by:

$$\hat{H}\phi_q(x_i, y_i, z_i) = -\frac{\hbar^2}{2m}[\sum_{n_1=-M}^{M} C_{n_1}\phi_q(x_i + n_1 h, y_i, z_i) + \sum_{n_2=-M}^{M} C_{n_2}\phi_q(x_i, y_i + n_2 h, z_i)$$

$$+ \sum_{n_3=-M}^{M} C_{n_3}\phi_q(x_i, y_i, z_i + n_3 h)] + [V_{ion}(x_i, y_i, z_i)$$

$$+ V_H(x_i, y_i, z_i) + V_{xc}(x_i, y_i, z_i)]\phi_q(x_i, y_i, z_i), \tag{4.11}$$

where $2M$ is order of finite difference, h is a lattice step, (x_i, y_i, z_i) is a point in discretized space. Coefficients C_{n_k} for a different orders of approximation are given in Table 4.1.

External potential also includes ionic potential V_{ion} that represents combined pseudopotential of nuclei and core electrons. Use of pseudopotentials dramatically

Table 4.1 Coefficients for uniform grid representation of Laplacian operator. N_{FD} is order of finite difference

N_{FD}								
2				1	-2	1		
4			$-\frac{1}{12}$	$\frac{4}{3}$	$-\frac{5}{2}$	$\frac{4}{3}$	$-\frac{1}{12}$	
6		$\frac{1}{90}$	$-\frac{3}{20}$	$\frac{3}{2}$	$-\frac{48}{18}$	$\frac{3}{2}$	$-\frac{3}{20}$	$\frac{1}{90}$
8	$-\frac{1}{560}$	$\frac{8}{315}$	$-\frac{1}{5}$	$\frac{8}{5}$	$-\frac{205}{72}$	$\frac{8}{5}$	$-\frac{1}{5}$	$\frac{8}{315}$ $-\frac{1}{560}$

increases efficiency of calculations. While local versions of pseudopotentials exist, the most efficient are fully separable pseudopotentials that have local and angular momentum dependent non-local parts [8]:

$$V_{ion} = \sum_{a=1}^{N_{nuc}} V_{loc,a} + \sum_{a=1}^{N_{nuc}} \sum_{l=1}^{l_{max}} \sum_{m=-l}^{l} c_{a,l,m} \hat{U}_{a,l,m} \hat{U}_{a,l,m}^{T} \tag{4.12}$$

Non-local vectors $\mathcal{U}_{a,l,m}$ are sparse and vanish outside of spherical regions surrounding atoms. The radii of these regions are atom dependent. $c_{a,l,m}$ are normalization constants.

4.4 The Real Time Algorithm

The algorithm contains three basic steps. These are the Hamiltonian operation $H[\rho]|\phi\rangle$ which is used recursively to construct the evolution operator U, solution of Poisson equation and calculation of the exchange-correlation potentials. The algorithm proceed as following. Starting with ground state orbitals ϕ_k at $t = 0$:

1. Construct evolution operator at time $t' = t + \delta t$ according to (4.10).
2. Apply evolution operator $U(t', t)|\phi_k(t)\rangle$.
3. Calculate density at time $t' = t + \delta t$: $\rho(t') = \sum \phi^*(t')\phi(t')$.
4. Solve Poisson equation for Hartree potential: $\nabla^2 V_H = -4\pi\rho(t')$.
5. Calculate exchange correlation potential $V_{xc}[\rho(t')]$.
6. Repeat steps 1–5 for the next time increment $t'' = t' + \delta t$.

The procedure stops when maximum number of steps have been reached. Various quantities typically are calculated at each time steps. These are the total energy, number of electrons and the observables of interest, such as total polarization.

References

1. C. Fiolhais, F. Nogueira, M. Marques (eds.), *A Primer in Density Functional Theory* (Springer, Berlin, 2003)
2. R. Dreizler, E. Gross, *Density Functional Theory: An Approach to the Quantum Many-Body Problem* (Springer, Berlin, 1990)

3. H. Eschrig, *Fundamentals of Density Functional Theory* (Eagle, Leipzig, 2003)
4. E. Runge, E. Gross, Phys. Rev. Lett. **52**, 997 (1984)
5. J.P. Perdew, A. Zunger, Phys. Rev. B **23**, 5048 (1981)
6. S.H. Vosko, L. Wilk, M. Nusair, Can. J. Phys. **58**, 1200 (1980)
7. J.R. Chelikowsky, N. Troullier, K. Wu, Y. Saad, Phys. Rev. B **50**, 11355 (1994)
8. Y. Saad, J. Chelikowsky, S. Shontz, SIAM Rev. **52**, 3 (2010)

Chapter 5
Response Functions from Real Time TDDFT

5.1 Method I: Reduction to a Linear System

The key characteristic of nth order non-linear optical response is that it supplies a contribution to total polarization that depends on nth power of electric field $\mathbf{E}^n$:

$$P_i = \sum_n \sum_{k...m} P^{(n)}_{ik...m}(\mathbf{E}^n). \tag{5.1}$$

The extraction procedure for $\chi^{(n)}$ in general would contain steps to decompose total polarization into sum of different orders (5.1) and then deduce $\chi^{(n)}$ from corresponding orders of nonlinear polarization $P^{(n)}$. The nth order response in frequency domain takes the following form:

$$P^{(n)}_{ik...m}(\omega) = \mathscr{K} \int \chi^{(n)}_{ik...m}\left(-\omega;\ \omega_1,\ \ldots,\ \omega - \sum_{j=1}^{n-1}\omega_j\right)$$

$$\times\ E_k(\omega_1)\ \ldots\ E_m\left(\omega - \sum_{j=1}^{n-1}\omega_j\right) d\omega_1 \ldots d\omega_{n-1}. \tag{5.2}$$

$\mathscr{K}$ is a factor that depends both on order and specific non-linear process. Tables 5.1 and 5.2 list these factors for several common processes.

The arguments of $\chi^{(n)}$ are customarily written as to yield a formal zero sum $-\omega + \omega_1 + \cdots + \omega_{n-1} + \omega - \sum_{j=1}^{n-1}\omega_j = 0$ to indicate (and enforce) the conservation of energy. Alternatively, a delta-function may be employed.[1]

---

[1] See Appendix B for details.

© The Author(s) 2014
V. Goncharov, *Non-Linear Optical Response in Atoms, Molecules and Clusters*,
SpringerBriefs in Electrical and Magnetic Properties of Atoms, Molecules, and Clusters,
DOI 10.1007/978-3-319-08320-9_5

Table 5.1 $\mathcal{K}$-factors for third order processes: $\mathcal{K} = \frac{D}{2^l (2\pi)^m}$

D	m	l	$(2\pi)^{-m}$	$-\omega_\sigma$	ω_1	ω_2	ω_3	$\mathcal{K}$	Process
1	0	1	1	0	0	0	0	1	Static
6	1	1	$\frac{1}{2\pi}$	0	0	$-\omega$	ω	$\frac{3}{2\pi}$	
3	0	0	1	$-\omega$	0	0	ω	3	Kerr
3	2	0	$\frac{1}{4\pi^2}$	$-\omega$	ω	ω	$-\omega$	$\frac{3}{4\pi^2}$	IDRI/TPA
1	2	0	$\frac{1}{4\pi^2}$	-3ω	ω	ω	ω	$\frac{1}{4\pi^2}$	THG
6	2	0	$\frac{1}{4\pi^2}$	$-\sum_i \omega_i$	ω_1	ω_2	ω_3	$\frac{3}{2\pi^2}$	$\omega_1 \neq \omega_2 \neq \omega_3$

D is number of non-equivalent permutations of inputs frequencies, m is one less than number of non-zero input frequencies, l is one if $\omega_\sigma = 0$, zero otherwise

Table 5.2 $\mathcal{K}$-factors for second order processes: $\mathcal{K} = \frac{D}{2^l (2\pi)^m}$

D	m	l	$(2\pi)^{-m}$	$-\omega_\sigma$	ω_1	ω_2	$\mathcal{K}$	Process
2	1	1	$\frac{1}{2\pi}$	0	$-\omega$	ω	$\frac{1}{2\pi}$	OR
2	0	0	1	$-\omega$	0	ω	2	Pokels
1	1	0	$\frac{1}{2\pi}$	-2ω	ω	ω	$\frac{1}{2\pi}$	SHG
2	1	0	$\frac{1}{2\pi}$	$-\sum_i \omega_i$	ω_1	ω_2	$\frac{1}{\pi}$	$\omega_1 \neq \omega_2$

D is a number of non-equivalent permutations of inputs frequencies, m is one less than number of non-zero input frequencies, l is one if $\omega_\sigma = 0$, zero otherwise

5.1.1 Time Dependent Density

The starting point is the modeling of interaction of electromagnetic field and a system of interest. The goal is to calculate a realistic density and polarization response to a model laser pulse. Time dependent density is obtained from time dependent Kohn-Sham orbitals that are propagated using Real-Time TDDFT method[2]:

$$\rho(\mathbf{r}, t) = \sum_k f(k)\phi_k^*(\mathbf{r}, t)\phi_k(\mathbf{r}, t), \tag{5.3}$$

where $f(k)$—occupation numbers. Hamiltonian $\hat{H}(t)$ contains time dependent potential representing electric dipole interaction of electrons with electric field of external electromagnetic wave.

$$V_{ext}(t) = -e\mathbf{E}(t) \cdot \mathbf{r} \tag{5.4}$$

with

$$\mathbf{E}(t) = (\lambda\hat{i} + \mu\hat{j} + \nu\hat{k}) \sin(\omega t) e^{-\frac{(t-t_0)^2}{\sigma^2}}, \tag{5.5}$$

---

[2] Additional details are presented in Chap. 3.

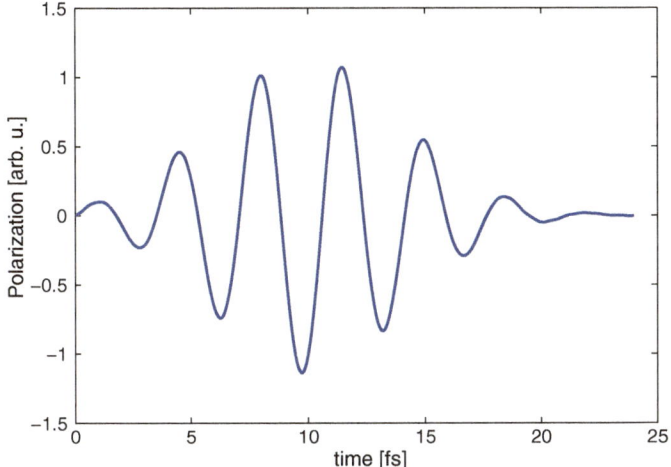

Fig. 5.1 Time dependent polarization $P^{tot}(t)$ of C_{60} fullerene propagated with PZ functional and external quasi-monochromatic field tuned to $\hbar\omega_0 = 1.17\,eV$

where $\lambda, \mu, \nu \in \mathbf{R}$ are amplitudes of corresponding Cartesian components of $\mathbf{E}$. Fourier transform of (5.5) could be obtained analytically, but keeping in mind a more general case we write it simply as

$$\mathbf{E}(\omega) = (\lambda\hat{i} + \mu\hat{j} + \nu\hat{k})g(\omega) \tag{5.6}$$

Total polarization is obtained from density and is time dependent:

$$\mathbf{P}^{tot}(t) = \int \mathbf{r}\delta\rho(\mathbf{r}, t)d^3r \tag{5.7}$$

An example of time dependent polarization for C_{60} fullerene is shown in Fig. 5.1. It is subsequently Fourier transformed:

$$\mathbf{P}^{tot}(\omega) = \int \mathbf{P}(t)e^{i\omega t}dt \tag{5.8}$$

Total polarization is a function of both frequency and strength of external electric field that is parametrized with (λ, μ, ν):

$$\mathbf{P}^{tot} = \mathbf{P}^{tot}(\omega, (\lambda, \mu, \nu)). \tag{5.9}$$

The quantities of interest are coefficients of Taylor expansion of $\mathbf{P}^{tot}(\omega, (\lambda, \mu, \nu))$ in respect to λ, μ, ν. The coefficients are found by converting several Taylor expansions of total polarization at different values of λ, μ, ν into a linear system.

5.1.2 Extraction in Case of Isotropic Symmetry

As an illustration of the method consider a system with spherical (isotropic) symmetry, and a single frequency external wave. The second order processes will vanish, and in cases of linear and third order responses there will be only one independent component of susceptibility.[3] Suppose that we are interested in calculating $\chi^{(3)}_{xxxx}$. One would set μ, ν to zero, and propagate wavefunction three times, each with different value of $\lambda_1, \lambda_2, \lambda_3$. Next, one calculates three corresponding polarizations $\{P^{tot}_x(\lambda_i), i = 1, 2, 3\}$:

$$P^{tot}_x(\omega, \lambda) = \int \int x \delta\rho(\mathbf{r}, t) e^{i\omega t} d^3r dt, \tag{5.10}$$

and writes down three polynomial expansions:

$$P^{tot}_x(\omega, \lambda_i) = a(\omega)\lambda_i + b(\omega)\lambda_i^2 + c(\omega)\lambda_i^3. \tag{5.11}$$

By introducing a matrix of electric field strength $\hat{\Theta}$:

$$\hat{\Theta} = \begin{pmatrix} \lambda_1, \lambda_1^2, \lambda_1^3 \\ \lambda_2, \lambda_2^2, \lambda_2^3 \\ \lambda_3, \lambda_3^2, \lambda_3^3 \end{pmatrix}, \tag{5.12}$$

as well as vectors of polarizations $\mathbf{P} = (P^{tot}_x(\omega, \lambda_1), P^{tot}_x(\omega, \lambda_2), P^{tot}_x(\omega, \lambda_3))$ and vector of coefficients $\mathbf{X} = (a(\omega), b(\omega), c(\omega))$, a set of equations of type (5.11) is written as

$$\mathbf{P} = \hat{\Theta}\mathbf{X}. \tag{5.13}$$

After solving Eq. (5.13) for $\mathbf{X}$, one gets $c(\omega)\lambda^3 = P^{(3)}_{xxxx}(\omega)$. Since $P^{(3)}_{xxxx}(\omega)$ is known, then $\chi^{(3)}_{xxxx}$ can be extracted from (5.2). $\chi^{(3)}$ appears in (5.2) in a non-local form. However, in case of quasi-monochromatic excitation the response is well localized in frequency domain, and $\chi^{(3)}$ could be pulled out of the integral[4] [1].

$$\mathcal{K} \int \chi^{(3)}_{xxxx}(-\omega; \omega'', \omega', \omega - \omega'' - \omega')\lambda^3 \, g(\omega') \, g(\omega'') \, g(\omega - \omega'' - \omega') \, d\omega' \, d\omega''$$

$$\sim \mathcal{K}\chi^{(3)}_{xxxx}(-\omega)\lambda^3 \int g(\omega') \, g(\omega'') \, g(\omega - \omega'' - \omega') \, d\omega' \, d\omega'' = \mathcal{K}\chi^{(3)}_{xxxx}(-\omega)\lambda^3 \mathcal{G}(\omega). \tag{5.14}$$

---

[3] In this case there exists the following relation between components of the third order susceptibility:
xxxx = yyyy = zzzz = 3xxyy = 3xxyy = 3yyxx = 3zzyy = 3zzxx.
[4] The locality is further discussed in the next section.

Table 5.3 Convolution integrals $\mathscr{G}$ for selected processes

Process	$\mathscr{G}(\omega)$
SHG	$\int_0^\infty g(\omega') \, g(2\omega - \omega') \, d\omega'$
OR	$\int_0^\infty g(\omega') g^*(\omega') d\omega'$
THG	$\int_0^\infty \int_0^\infty g(\omega') g(\omega'') g(3\omega - \omega' - \omega'') d\omega' d\omega''$
IDRI/2-photon absorption	$\int_0^\infty \int_0^\infty g^*(\omega') g(\omega'') g(\omega + \omega' - \omega'') d\omega' d\omega''$

The convolution $\mathscr{G}(\omega)$ depends on shape of external field and on the type of process, and is shown for four common processes in Table 5.3. The second hyperpolarizability becomes:

$$\chi_{xxxx}^{(3)})(\omega) = \frac{c(\omega)}{\mathscr{K}\,\mathscr{G}(\omega)}. \tag{5.15}$$

Selecting appropriate $\mathscr{K}$ and $\mathscr{G}$ provides information about THG, IDRI and 2-photon absorption.

5.1.3 Generalization

Off diagonal components of $\chi^{(n)}$ require more than one field. It depends on the symmetry of molecule (or crystal) which components are necessary to calculate. In case of molecules, hyperpolarizabilities are typically spatially averaged to account for the random orientation. This allows us to compare the calculated and experimental results obtained from the gas phase. Tables 5.4 and 5.5 list components and selection of fields that are needed to obtain spatially averaged $\chi_{||}^{(2)}$ and $\chi_{||}^{(3)}$ in case when the molecular symmetry is neglected, or when molecule does not have any symmetry:

$$\chi^{(1)} = \frac{1}{3} \sum_{i=x,y,z} \chi_{ii}^{(1)} \tag{5.16}$$

$$\chi^{(2)} = \frac{1}{5} \sum_{i=x,y,z} \left(\chi_{zii}^{(2)} + \chi_{izi}^{(2)} + \chi_{iiz}^{(2)} \right) \tag{5.17}$$

Table 5.4 Tensor components needed for evaluation of spatially averaged hyperpolarizabilities

Degeneracy	Component	Field direction
1	zxx	x
1	zxx	y
1	zzz	z
2	xxz	x, z
2	yyz	y, z

Table 5.5 Tensor components needed for evaluation of spatially averaged second hyperpolarizabilities

Degeneracy	Component	Field direction
1	xxxx	x
1	yyyy	y
1	zzzz	z
3	xxyy	x, y
3	yyxx	x, y
3	yyzz	y, z
3	zzyy	y, z
3	xxzz	x, z
3	zzxx	x, z

$$\chi^{(3)} = \frac{1}{15} \sum_{i=x,y,z} \sum_{j=x,y,z} \left(\chi_{iijj}^{(3)} + \chi_{ijij}^{(3)} + \chi_{ijji}^{(3)} \right). \tag{5.18}$$

Generalization for off-diagonal components is simple. For compactness we re-label electric fields as

$$E_j(\omega) = \varepsilon_j \, g(\omega),$$

vector of coefficients as

$$\mathbf{X} = (a_{11}^{(1)}, a_{12}^{(1)}, a_{13}^{(1)}, \dots, a_{111}^{(2)}, a_{112}^{(2)}, a_{113}^{(2)}, \dots, a_{1111}^{(3)}, a_{1112}^{(3)}, a_{1113}^{(3)}, \dots), \tag{5.19}$$

vector of polarizations as

$$\mathbf{P} = (P_1(\omega, \mathbf{E}(1)), P_2(\omega, \mathbf{E}(1)), P_3(\omega, \mathbf{E}(1)), \dots, P_1(\omega, \mathbf{E}(\zeta)), P_2(\omega, \mathbf{E}(\zeta)), P_3(\omega, \mathbf{E}(\zeta))), \tag{5.20}$$

and matrix of field strengths as

$$\hat{\Theta} = \begin{pmatrix} \varepsilon_1(1) \ \varepsilon_2(1) \ \varepsilon_3(1) \ \dots \ \varepsilon_1^2(1) \ \varepsilon_1\varepsilon_2(1) \ \varepsilon_1\varepsilon_3(1) \ \dots \ \varepsilon_1(1)^3 \ \varepsilon_1(1)^2\varepsilon_2(1) \ \dots \\ \dots \quad \dots \quad \dots \quad \dots\dots \quad \dots \quad \dots \quad \dots\dots \quad \dots \quad \dots \\ \varepsilon_1(\zeta) \ \varepsilon_2(\zeta) \ \varepsilon_3(\zeta) \ \dots \ \varepsilon_1^2(\zeta) \ \varepsilon_1\varepsilon_2(\zeta) \ \varepsilon_1\varepsilon_3(\zeta) \ \dots \ \varepsilon_1(\zeta)^3 \ \varepsilon_1(\zeta)^2\varepsilon_2(\zeta) \ \dots \end{pmatrix}, \tag{5.21}$$

where $k, l, m = 0, 1, 2, \dots, \eta = 1, 2, \dots, \zeta$ is a index labeling set of field amplitudes. In addition, $\hat{\Theta}$ is subject to

$$\det \left| \hat{\Theta} \right| \neq 0. \tag{5.22}$$

Then, similarly to (5.15), the response function $\hat{\chi}^{(n)}$ is:

$$\chi_{ik...m}^{(n)}(-\omega) = \frac{a_{ik...m}^{(n)}(\omega)}{\mathcal{K}\,\mathcal{G}(\omega)}. \tag{5.23}$$

Variants of the above method include "fitting" in either frequency or time domain. In these cases one sets up an overdetermined linear system, where $dim(\mathbf{P}) > dim(\mathbf{X})$, and formulates the problem as a linear least squares problem, where $\mathbf{X}$ is sought as a minimum of

$$\|\hat{\Theta}\,\mathbf{X} - \mathbf{P}(\omega)\|_2. \tag{5.24}$$

However, the direct solution of (5.13) is preferable to "fitting", because it requires less data. Other approaches for extraction of $\chi^{(k)}$ exist, in particular numerical differentiation in frequency domain [1].

5.2 Locality of Nonlinear Response Under Monochromatic Excitation

The possibility of factoring $\chi^{(n)}$ out of integral (5.2) depends on how rapidly $\chi^{(n)}$ changes in a range of frequencies that makes contribution to the convolution integral $\mathcal{G}$ appearing in (5.14).[5] We do not make any assumptions on how $\chi^{(n)}$ behaves, except that it is continuous in some frequency region $\Omega_{n-1}^\chi \in \mathbf{R}^{n-1}$, $\Omega_{n-1}^\chi = [\omega_1', \omega_1] \otimes [\omega_2', \omega_2''] \otimes \cdots$. However, we are free to define the shape of external electric fields. Setting fields to delta function would collapse the convolution integral entirely, which is equivalent to using purely monochromatic excitation. It is not practically possible to simulate a purely monochromatic pulse within RT-TDDFT. Instead one may use a Gaussian shaped field as a model of quasi monochromatic laser pulse.

For quasi monochromatic excitation with frequency ω_0 the absolute value of $|E(\omega)|$ asymptotically decreases outside of a small interval of frequencies centered at ω_0. If one sets a threshold $\varepsilon_M > 0$, then one may say that $E(\omega')$ is localized within interval Ω_1 if $\forall \omega' \in \Omega_1 \Rightarrow |E(\omega')| \leq \varepsilon_M$. For example, $E(\omega) = \lambda \int \sin(\omega_0 t) e^{-\frac{(t-t_0)^2}{4\sigma^2}} e^{i\omega t} dt$ is localized at $\Omega_1 = [\omega_0 - \Delta(\sigma, \varepsilon_M), \omega_0 + \Delta(\sigma, \varepsilon_M)]$, where

$$\Delta(\sigma, \varepsilon_M) = \frac{1}{|\sigma|}\sqrt{|\ln \frac{\varepsilon_M}{2\lambda\sigma\sqrt{\pi}}|}. \tag{5.25}$$

The products of electric fields $E(\omega')E(2\omega_0 - \omega')$ and $E(\omega')E^*(\omega')$ have the same localization as $E(\omega)$, while $E(\omega')E(\omega'')E(3\omega_0 - \omega' - \omega'')$ and $E(\omega')E^*(\omega'')E(\omega_0 - \omega' + \omega'')$ have $\Omega_2 \approx \Omega_1 \otimes \Omega_1$. Moreover, for higher harmonic generation of nth order $\Omega_{n-1} \approx \Omega_1 \otimes \Omega_1 \otimes \Omega_1 \cdots \approx [\Omega_1]^{n-1}$. The size of Ω_n is controlled by parameter σ, and could always be made smaller. Thus, as long as $\chi^{(n)}(\omega_1, \omega_2, \ldots, \omega_n)$ is continuous on Ω_{n-1}, one can adjust σ to make change in $\chi^{(n)}$ smaller then a threshold ε_χ:

---

[5] See examples of $\mathcal{G}$ in Table 5.3.

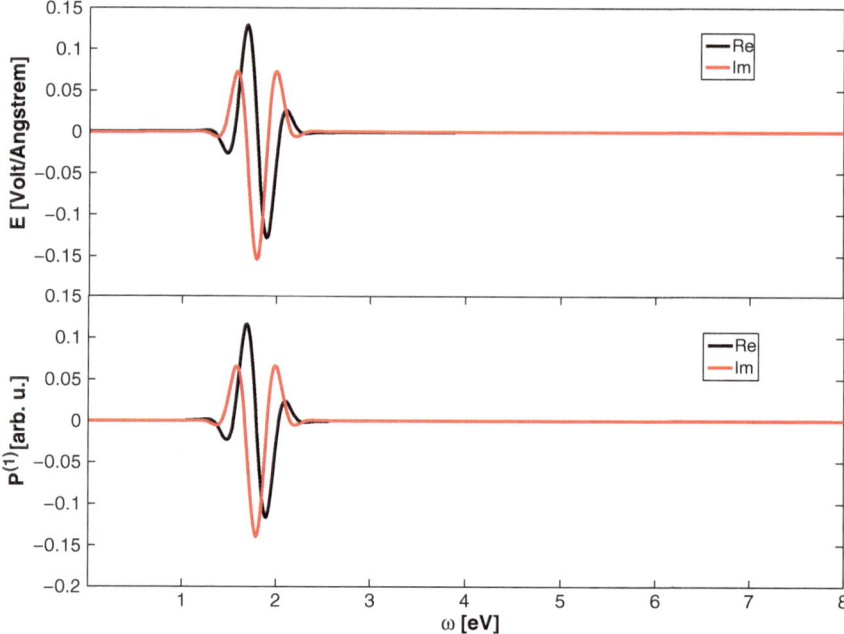

Fig. 5.2 Linear polarization response to quasi monochromatic excitation at $\hbar\omega = 1.79$ eV. On the *top* is electric field $E(\omega)$. At the *bottom* is second order nonlinear polarization $P^{(1)}(\omega)$. Real part is in *black*, imaginary part is *red*. Data is shown for H_2O molecule propagated using LB94 functional [2]

$$|\chi^{(n)}(\omega_1, \omega_2 \ldots, \omega_n) - \chi^{(n)}(\omega_1', \omega_2' \ldots, \omega_n')| < \varepsilon_\chi \quad \forall \omega_i, \omega_i' \in \Omega_{n-1}. \quad (5.26)$$

Therefore, if $\chi^{(n)}$ is continuous on Ω_{n-1}, then it can be made local (factorable from the integral) on $\Omega_{n-1}^\chi \subset \Omega_{n-1}$. Presence of finite number of poles in Ω_{n-1} breaks Ω_{n-1} into regions Ω_{n-1}', where locality of $\chi^{(n)}$ can be re-established: $\Omega_{n-1}^{\chi'} \subset \Omega_{n-1}'$. For example, in case of a second order process presence of a resonance ω_R on Ω_1 will break it into two frequency intervals. Calculations of $\chi^{(2)}$ then will proceed by approaching ω_R from left $\omega_R - \delta$ and right $\omega_R - \delta$ and progressively increasing σ. Now, we turn to the analysis of polarization response in frequency domain under quasi-monochromatic excitation. We use Gaussian shaped external electric field pulse tuned to $\hbar\omega_0 = 1.79$ eV to probe response of H_2O molecule.[6] At this frequency the optical interaction is lossless and dispersionless. The molecule was propagated and total polarization was calculated as described above. Total polarization was decomposed into the sum of first three orders using Method I (a.k.a. Linear Reduction method) described in the previous section. Figure 5.2 shows that linear polarizability and electric field have practically identical shapes. This is only possible

---

[6] Similar results are obtained for other small organic molecules such as carbon monoxide, hydrogen fluoride etc.

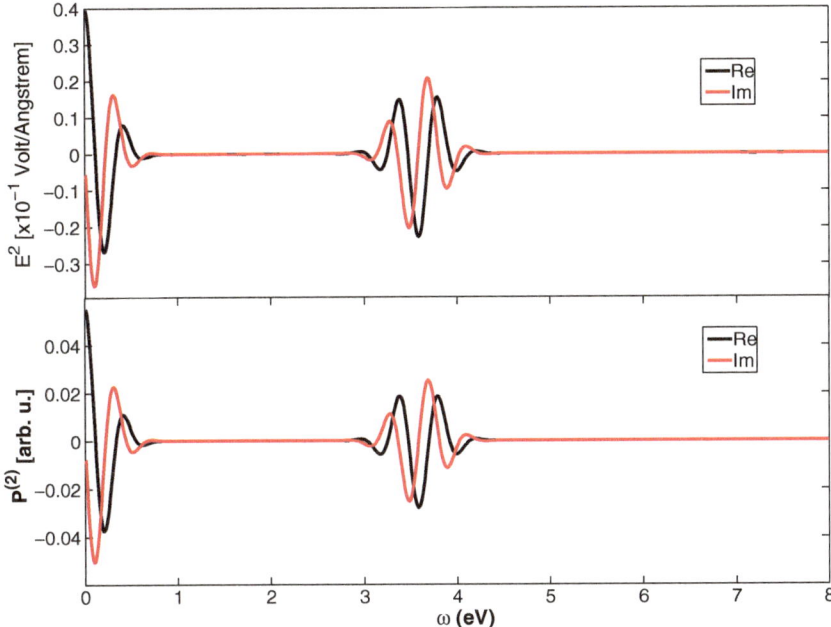

Fig. 5.3 Second order polarization response to quasi-monochromatic excitation at $\hbar\omega = 1.79\,\text{eV}$. On the *top* is square of electric field $E^2(\omega)$. At the *bottom* is second order nonlinear polarization $P^{(2)}(\omega)$. Real part is in *black*, imaginary part is *red*. Data is shown for H_2O molecule propagated using LB94 functional [2]

if susceptibility is a real constant, since in frequency domain linear polarization is proportional to both electric field and generally frequency dependent susceptibility: $P(\omega) = \chi^{(1)}(\omega)E(\omega)$. Indeed this is the case in the small frequency interval near 1.79 eV, because interaction is lossless and dispersionless. Polarization vanishes outside of 1.1–2.44 eV range. This agrees with 0.1 % localization range of ± 0.95 eV. Now we turn to higher orders. Figure 5.3 shows second order polarization for H_2O molecule. Shape of second order polarization is identical to square of applied electric field $E^2(\omega)$. This implies that not only $\chi^{(2)}$ is a real constant, but also that we have similar relation as in linear case: $P^{(2)}(\omega) = D^{(2)}\chi^{(2)}(\omega)E(\omega)E(\omega)$, where $D^{(2)}$ is degeneracy factor. Same argument holds for $\chi^{(3)}$. From Fig. 5.4 we deduce that $P^{(3)}(\omega) = D^{(3)}\chi^{(3)}(\omega)E(\omega)E(\omega)E(\omega)$. We see that generally non-local character of relationship between nonlinear polarization and corresponding susceptibility (5.2) is reducing to a simple product of nth power of electric field and nth order susceptibility. This means that under sufficiently narrow quasimonochromatic excitation polarization response resembles response under mononochromaric excitation. Then in (5.2) we can make replacement $E(\omega_i) \rightarrow \delta(\omega_i - \omega_0)$ and reduce integral to a simple product of electric fields and corresponding susceptibility. This is one of the two practically important features of quasimonochromatic probes. Another one may

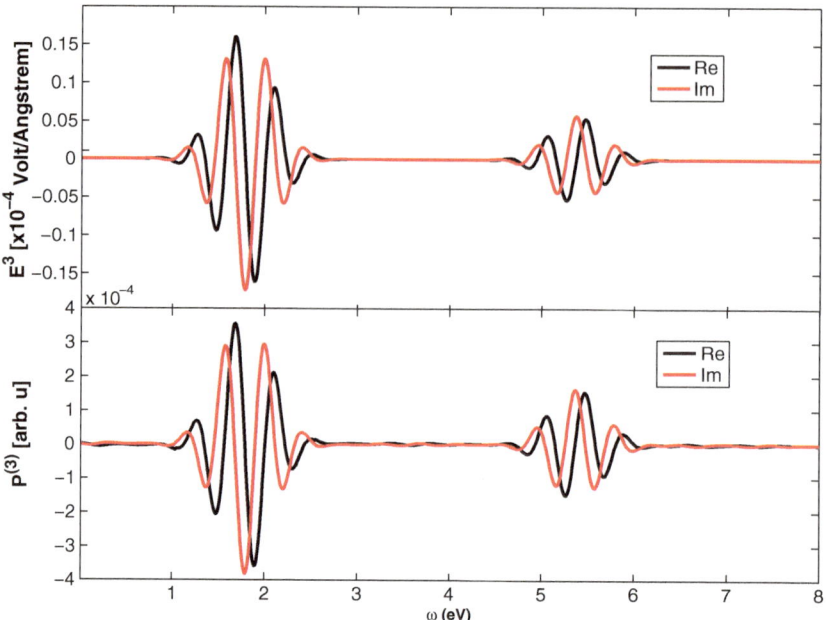

Fig. 5.4 Third order polarization response to quasi-monochromatic excitation at $\hbar\omega = 1.79\,\text{eV}$. On the *top* is cube of electric field $E^3(\omega)$. At the *bottom* is third order nonlinear polarization $P^{(3)}(\omega)$. Real part is in *black*, imaginary part is *red*. Data is shown for xxzz component of H_2O molecule propagated using LB94 functional

be easily inferred from Fig. 5.5. It is clear that second and third order responses are localized within *non-intersecting* frequency intervals. Therefore, under assumption that there is only second and third responses present, total polarization decomposes trivially into sum of second and third orders for SHG, THG and OR processes. This is the basis for Direct Evaluation Method for obtaining susceptibilities described below.

5.3 Method II: Direct Evaluation

Because under quasi-monochromatic excitation odd and even orders of nonlinear optical response are resolved, we may write:

$$P^{tot}(0) = P^{(2)}(0), \tag{5.27}$$

$$P^{tot}(2\omega) = P^{(2)}(2\omega), \tag{5.28}$$

$$P^{tot}(3\omega) = P^{(3)}(3\omega). \tag{5.29}$$

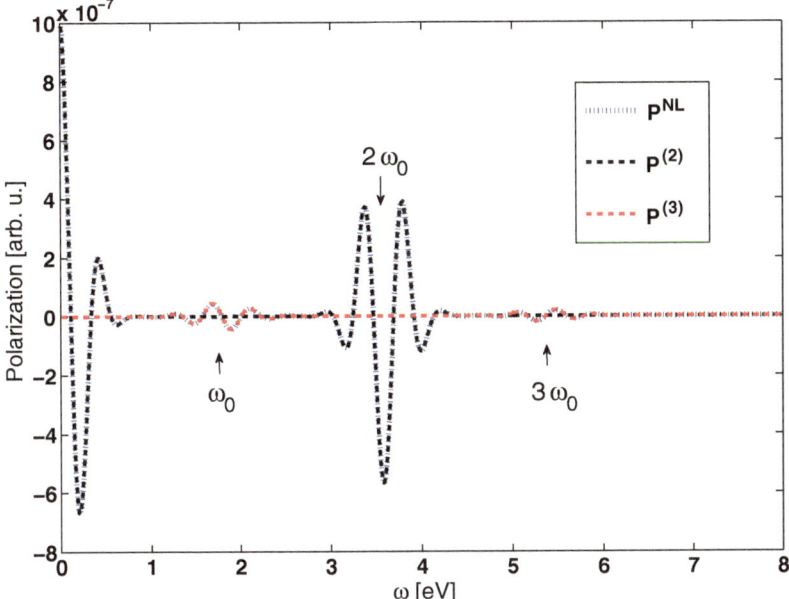

Fig. 5.5 Decomposition of real part of nonlinear polarization $P^{NL} = P^{tot} - P^{(1)}$ into second ($P^{(2)}(\omega)$) and third order ($P^{(3)}(\omega)$) polarizations in frequency space. The different orders of polarization are resolved in the frequency domain. Data is shown for xxz component of H_2O molecule propagated using LB94 functional [2]

These relations are valid only if the higher order responses are negligible. This condition can be achieved in most of practical situations by selecting the appropriate amplitude of perturbing electric field. This means that, when for OR, SHG and THG decomposition $P^{tot} = \sum P^{(n)}$ happens automatically in frequency domain, we can extract diagonal elements of $\chi^{(2)}$ and $\chi^{(3)}$ for these processes from a single propagation. If field has one component $\mathbf{E} = (E_x, 0, 0)$, then:

$$P_i^{tot} = \sum_{jk} D\chi_{ijk}^{(2)} E_j E_k = D\chi_{ixx}^{(2)} E_x^2. \qquad (5.30)$$

Diagonal components are found from:

$$\chi_{ixx}^{(2)}(\omega) = \frac{P_i^{tot}(\omega)}{DE_x^2(\omega)}. \qquad (5.31)$$

After diagonal components are found, one sets the electric fields to $\mathbf{E}(\omega) = (E_x(\omega), E_y(\omega), 0)$:

$$P_i^{tot} = \sum_{jk} D\chi_{ijk}^{(2)} E_j E_k = \mathscr{K} \chi_{ixx}^{(2)} E_x^2 + D\chi_{ixx}^{(2)} E_y^2 + 2D\chi_{ixy}^{(2)} E_x E_y. \qquad (5.32)$$

Setting $E_x(\omega) = E_y(\omega) = E(\omega)$ and calculating $\chi_{ixx}^{(2)}$ and $\chi_{iyy}^{(2)}$ independently one gets for $\chi_{ixy}^{(2)}$:

$$\chi_{ixy}^{(2)}(\omega) = \frac{1}{2}\left(\frac{P_i^{tot}(\omega)}{DE^2(\omega)} - \chi_{ixx}^{(2)}(\omega) - \chi_{iyy}^{(2)}(\omega)\right). \qquad (5.33)$$

Thus, from five propagations one obtains five components necessary for averaging (5.17). Equation (5.33) is valid for SHG and OR processes.

Similarly for THG:

$$P_i^{tot}(3\omega) = \sum_{jkl} D\chi_{ijkl}^{(3)} E_j E_k E_l = D\chi_{ixxx}^{(3)} E_x^3(3\omega). \qquad (5.34)$$

Diagonal components are

$$\chi_{ixxx}^{(3)}(3\omega) = \frac{P_i^{tot}(3\omega)}{DE_x^3(3\omega)}, \qquad (5.35)$$

and off diagonals are:

$$\chi_{iyxx}^{(3)}(3\omega) = \frac{1}{6}\left(\frac{P_i^{tot}(3\omega)}{DE^3(3\omega)} - \chi_{ixxx}^{(3)}(3\omega) - \chi_{iyyy}^{(3)}(3\omega)\right). \qquad (5.36)$$

Because linear and third order responses overlay at fundamental frequency ω_0, IDRI/TPA requires more then one propagation. For diagonal components we have:

$$\begin{cases} P^{tot}(\omega) = D\chi^{(3)}(\omega)E^2(\omega)E^*(\omega) + \chi^{(1)}(\omega)E(\omega) \\ P^{tot'}(\omega) = D\chi^{(3)}(\omega)E^{2'}(\omega)E^{*'}(\omega) + \chi^{(1)}(\omega)E'(\omega) \end{cases}. \qquad (5.37)$$

P^{tot} is obtained from propagation under field $E(\omega)$, and $P^{tot'}$ is obtained from propagation under field $E'(\omega)$. Diagonal components are obtained from

$$\chi_{ixxx}^{(3)}(\omega) = \frac{P_i^{tot}(\omega)E'(\omega) - P_i^{tot'}(\omega)E(\omega)}{D(E^3(\omega)E'(\omega) - E^{3'}(\omega)E(\omega))}. \qquad (5.38)$$

Similar considerations apply to off-diagonal components.

The Direct method is significantly more efficient than any other real-time method. In case of diagonal components only three propagations are needed, while the Linear Reduction method requires nine. Spatially averaged $\beta_\parallel$ requires five propagations in case of Direct method *versus* twenty five in case of Linear Reduction, making approximately fivefold reduction of total calculation time.

5.4 The Role of the Convolution Integral $\mathscr{G}(\omega)$ in Calculations of Response Functions

One may not fail to notice that convolution integral $\mathscr{G}(\omega)$ disappears from expression for nonlinear susceptibility Eqs. (5.33), (5.36), and (5.38). This is only possible if $\mathscr{G}(\omega)$ is either a real or an imaginary constant. This is precisely the case for a Gaussian pulse. $\mathscr{G}(\omega)$ for SHG process is shown in Fig. 5.6. It is a real constant within a range of frequencies relevant for calculation of SHG response. The role of $\mathscr{G}(\omega)$ is to account for the finite width of quasi-monochromatic excitation. Therefore it can depend on the width of the pulse but not on the frequency of excitation.

In case of THG $\mathscr{G}(\omega)$ is purely imaginary.

5.5 Method III: Differentiation in Frequency Domain

The fact that under quasi-monochromatic excitation polarization response is local in frequency domain allows one to calculate Taylor expansion for total polarization by direct differentiation. In this case, the partial derivatives are calculated by using finite differences. The following simple argument shows equivalence between linear decomposition and numerical differentiation in frequency domain. If we know polarization at specific frequency $p_i(\omega)$, we can write it in the following form:

Fig. 5.6 The convolution integral $\mathscr{G}(\omega)$ in case of SHG process. It is a constant everywhere except in a small interval near zero, where integration breaks down

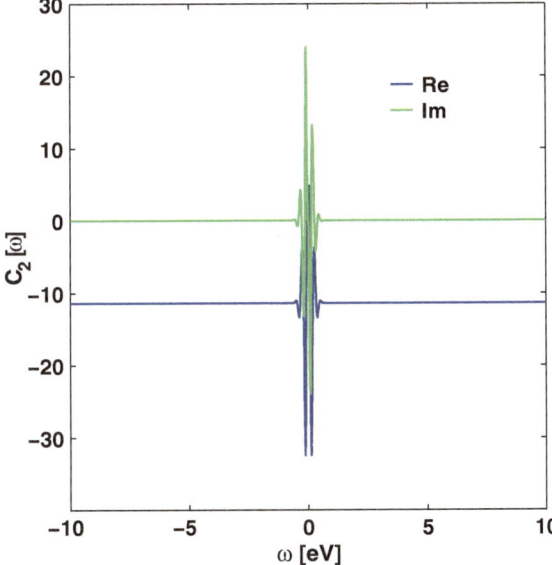

$$p_i(\omega) = p_i^{(1)}(\omega) + p_{ijk}^{(2)}(\omega)\,\hat{e}^j\,\hat{e}^k + p_{ijkl}^{(3)}(\omega)\,\hat{e}^j\,\hat{e}^k\,\hat{e}^l + \cdots, \qquad (5.39)$$

where $\hat{e}^j$—are Cartesian orthogonal unit vectors. We can also write it as a formal Taylor expansion:

$$p_i(\omega) = \frac{\partial p_i(\omega)}{\partial E^j}\,\delta E_j + \frac{1}{2!}\frac{\partial^2 P_i(\omega)}{\partial E^j \partial E^k}\,\delta E_j\,\delta E_k + \frac{1}{3!}\frac{\partial^3 P_i(\omega)}{\partial E^j \partial E^k \partial E^l}\,\delta E_j\,\delta E_k\,\delta E_l + \cdots$$
$$(5.40)$$

Or we can write it as a polynomial in δE_k where coefficients $a_{ijk\ldots}^{(n)}$ are obtained by fitting a set of $\{p_i(\omega)\}$ computed at different fields:

$$p_i(\omega) = a_{ij}^{(1)}\,\delta E_j + a_{ijk}^{(2)}\,\delta E_j\,\delta E_k + a_{ijkl}^{(3)}\,\delta E_j\,\delta E_k\,\delta E_l + \cdots. \qquad (5.41)$$

Above we had used Einstein summation notation, and no summation will be applied below. Now, suppose we are interested in a specific component of the second order susceptibility $\chi_{ijk}^{(2)}$. By comparing second order terms in Eqs. (5.39), (5.40), and (5.41) we get:

$$\frac{1}{2\pi}\int \chi_{ijk}^{(2)}(-\omega;\omega',\omega-\omega')E_j(\omega')E_k(\omega-\omega')d\omega' = \frac{(2-\delta_{jk})}{2}\frac{\partial^2 p_i(\omega)}{\partial E^j \partial E^k}\,\delta E_j\,\delta E_k$$
$$= a_{ijk}^{(2)}\,\delta E_j\,\delta E_k = p_{ijk}^{(2)}(\omega).$$
$$(5.42)$$

Next, we factor out the amplitude of the electric field and write it as a product with unity normalized function $f(\omega)$ that contains frequency dependence:

$$E_j(\omega') = \delta E_j\,f(\omega').$$

In general case one has to solve an integral equation similar to (5.2), but here, for quasi-monochromatic fields we impose locality on $\chi_{ijk}^{(2)}$ and write:

$$2\pi\,p_{ijk}^{(2)}(\omega) = \chi_{ijk}^{(2)}(-\omega)\,\delta E_j\,\delta E_k \int f(\omega')f(\omega-\omega')\,d\omega'. \qquad (5.43)$$

Finally, we get the representations of $\chi_{ijk}^{(2)}$ as a fitting coefficient (middle), or a partial derivative (right):

$$\chi_{ijk}^{(2)}(-\omega) = \frac{2\pi\,a_{ijk}^{(2)}}{\int f(\omega')f(\omega-\omega')\,d\omega'} = \frac{(2-\delta_{jk})\,\pi}{\int f(\omega')f(\omega-\omega')\,d\omega'}\frac{\partial^2 p_i(\omega)}{\partial E^j \partial E^k}.$$
$$(5.44)$$

Thus, when the non-locality of $\chi^{(n)}$ in (5.2) could be neglected, the finite-difference method could be used directly in the frequency space. Practical calculations show

Table 5.6 Comparison between methods for HF

Method	$\chi_{zxx}^{(2)}$	$\chi_{zzz}^{(2)}$	$\chi_{\parallel}^{(2)}$
Frequency domain	−2.89253973358058	−11.6003281577394	−10.4312445749404
Time domain	−2.89253973358058	−11.6003281577394	−10.4312445749404
Finite difference	−2.89253980559773	−11.6003278763701	−10.4312444925393

For individual components, the numerical discrepancy occurs at 10^{-6} a.u. level. This agreement is typical and had served as an additional quality check for the data presented in this work

good agreement between the methods. Comparison between these methods can be found in Table 5.6.

5.6 Conclusion

The three methods described in this chapter have different accuracy, reliability and computational cost. The computational cost is dominated by the propagation of the wavefunction, and the number of propagations used by the method is a measure of its cost. Direct Evaluation is particularly suitable for large systems, because it needs as little as one propagation of wavefunction. It is also the least accurate as it may not distinguish between different orders of response, because of the limited number of propagations. The Linear Reduction method is optimal in terms of quality and computational cost. Each propagation corresponds to a term in the polynomial expansion of the total polarization (5.41). The accuracy is increased by increasing the number of propagations. The limitation of the method is its computational cost, which becomes important in case of large systems. The "fitting" methods previously used [3] are inherently less efficient than the Linear Reduction, because there are always more propagations than terms in the polynomial expansion. In other words, some propagations are "wasted", because they do not contribute to the increase in accuracy.

References

1. V.A. Goncharov, K. Varga, J. Chem. Phys. **137**, 094111 (2012)
2. R. van Leeuwen, E.J. Baerends, Phys. Rev. A **49**, 2421 (1994)
3. Y. Takimoto, F.D. Vila, J.J. Rehr, J. Chem. Phys. **127**, 154114 (2007)

Chapter 6
Nonlinear Response in Atoms, Molecules and Clusters

6.1 Numerical Considerations

There are two groups of factors that determine quality of real time real space TDDFT simulations. The first group includes the size of simulation cell, grid step, exchange-correlation functional, and convergence of forces and energy in ground state. The size of cell has to be large enough that at any step of calculations density at the periphery of the cell is close to zero. When this condition is met, size of the box does not affect calculated values of susceptibilities. Since total number of operations depends cubically on the size, it is important to choose the optimal size. Table 6.1 shows data for HF molecule. Supercell is a cube with side L. The susceptibilities show little dependence on L, because $L = 10\,\text{Å}$ is sufficient for this small molecule. For the calculations presented in this Chapter 14–$20\,\text{Å}$ cell was used. The second parameter is grid step. In Table 6.2 we hold $L = 14\,\text{Å}$ and vary the grid step. $\Delta x = 0.25\,\text{Å}$ is an acceptable choice for a grid step for a variety of molecules and atoms.[1] The LDA functionals PZ and VWN [2, 3] are a good first choice, although as practice shows they frequently give overestimated hyperpolarizabilities [1]. Among GGA functionals, LB94 [4] frequently gives better agreement with experiment than LDA. However, it may not conserve energy, tends to yield wrong HOMO-LUMO gap, and may affect stability of calculations.[2] Poorly converged ground state results in unphysical oscillations of polarizability. Better than 0.01 eV convergence in single particle energies is expected.

The second group of factors controlling quality of simulations defines the stability and fidelity of Real Time Evolution. Among this group are size of time step Δt, total simulation time and maximum strength of applied electric field. The stability of the propagation is critically dependent on the size of time step Δt. It is bounded by the

---

[1] It depends on implementation of the pseudopotentials. But once convergence criteria in respect to grid step are established for a particulate set of pseudopotentials, grid step doesn't have to be frequently adjusted.

[2] LB94 is susceptible to numerical instabilities because it calculates asymptotic Coulomb tail from density gradient in the regions of near zero density.

© The Author(s) 2014
V. Goncharov, *Non-Linear Optical Response in Atoms, Molecules and Clusters*,
SpringerBriefs in Electrical and Magnetic Properties of Atoms, Molecules, and Clusters,
DOI 10.1007/978-3-319-08320-9_6

Table 6.1 Dependence of $\chi^{(2)}$ on the size of simulation cell for HF. The grid step was kept at $0.25\,\text{Å}$

L(Å)	$\chi_{zxx}^{(2)}$	$\chi_{zzz}^{(2)}$	$\chi_{\parallel}^{(2)}$
10	−2.90	−11.62	−10.45
12	−2.90	−11.62	−10.45
14	−2.89	−11.60	−10.43
16	−2.89	−11.60	−10.43

Table 6.2 Dependence of $\chi^{(2)}$ on grid spacing for HF. Size of simulation cell was kept at $14\,\text{Å}$

Δx (Å)	$\chi_{zxx}^{(2)}$	$\chi_{zzz}^{(2)}$	$\chi_{\parallel}^{(2)}$
0.225	−2.84	−12.77	−11.07
0.250	−2.97	−11.86	−10.68
0.275	−4.98	−8.85	−11.29
0.300	−27.95	−17.03	−43.76

Table 6.3 Dependence of $\chi^{(2)}$ on the size of time step dt for HF. N_t—is the total number of steps. The total simulation time was kept at $26.21\,(\hbar/\text{eV})$

dt ($\hbar$/eV)	N_t	$\chi_{zxx}^{(2)}$	$\chi_{zzz}^{(2)}$	$\chi_{\parallel}^{(2)}$
10^{-3}	26214	−2.89	−11.60	−10.43
1.6×10^{-3}	16384	−2.89	−11.60	−10.43
2.0×10^{-3}	13108	−2.89	−11.60	−10.43
2.62×10^{-3}	10000	N/D	N/D	N/D

following expression [1]

$$0 < \Delta t < \sqrt{\frac{2}{9}m(\Delta x)^2}, \tag{6.1}$$

where m is electron mass. The error in wavefunction at each step is $\sim \mathcal{O}((\frac{\Delta t\,|E(t)|}{\hbar})^5)$, and in principle could be matched to machine precision by choice of Δt and the amplitude of external field E_{max}. Choosing $\Delta t \sim 10^{-3}$ fs ensures stable propagation for about 10^5 steps for majority of systems. When a small enough step is chosen to provide the computational stability through entire simulation, the results do not appear to vary with the size of the time step (see Table 6.3).

The choice of the strength of electric field is also important. On one hand the field has to be strong enough to elicit a robust nonlinear response of desirable order. On another, it should not be strong enough to excite higher order responses. If one wants to use direct evaluation method, then the highest response should be not higher than third order response. This makes the choice of field strength molecule dependent. We had used $E \sim 0.013\,(\text{V/Å})$ as a starting point, and then repeated calculations with increased field, occasionally as high as $1.0\,(\text{V/Å})$ (where most of the molecules undergo Coulomb explosion). Then we chose the region of the field where the response functions show least field dependence. The field range $E \sim 0.013$–$0.05\,(\text{V/Å})$ is satisfactory for all cases we have tested so far.

Table 6.5 shows the dependence of the first hyperpolarizability on the strength of the electric field for the CO molecule. An eightfold increase in the electric field

Table 6.4 Second order susceptibilities $\chi_{||}^{(2)}(-2\omega; \omega, \omega)$

Molecule	$|\mu|$	$\hbar\omega$	$\chi_{||}^{(2)}$	Exp.	GF	[5Z4P]
CO	0.116	1.79	35.49	30.2 ± 3.2	35.48	33.24
		1.96	37.07		36.89	34.70
H_2O	1.953	1.79	-35.45	-22.2 ± 0.9	-35.36	-28.90
		1.96	-38.13		-37.83	-30.9
HF	1.908	1.79	-10.39	-11.0 ± 1.0	-11.06	-10.58 *
		1.96	-10.65		-11.42	-10.93 *
H_2S	1.075	1.79	-32.39	-10.1 ± 2.1	-32.48	N/A
NH_3	1.585	1.79	-120.58	-48.9 ± 1.2	-119.9	N/A

The energy, $\hbar\omega$, is in eV; the calculated permanent dipole moment, $|\mu|$, is in (Debye); and $\chi^{(2)}$ is in atomic units. The experimental data is taken from [6]. GF ($\chi_{||}^{(2)}$) stands for the results obtained using method of Iwata and Yabana [1], [5Z4P] is from [9] and is calculated by using 5Z4P basis. * denotes data from [7]

Table 6.5 Dependence of $\chi^{(2)}$ on ΔE for CO. The field strength is given in (V/Å). Eight-fold increase in the field strength results in a $\sim 0.31\%$ decrease in the first hyper-polarizability

$|\Delta E_\alpha|$	$\chi_{zxx}^{(2)}$	$\chi_{zzz}^{(2)}$	$\chi_{||}^{(2)}$
0.013	10.74	38.46	35.97
0.026	10.72	38.43	35.92
0.053	10.71	38.43	35.90
0.106	10.69	38.38	35.86

Table 6.6 Dependence of $\chi^{(3)}$ on ΔE for N_2

$|\Delta E_\alpha|$	$\chi_{xxxx}^{(3)}$	$\chi_{zzzz}^{(3)}$	$\chi_{xxzz}^{(3)}$	$\chi_{||}^{(3)}$
0.013	1269	1655	380	1312
0.026	1220	1788	373	1307
0.053	1263	1680	386	1318
0.106	1397	2085	470	1539

The field strength is given in (V/Å). Eight-fold increase in the field strength results in a $\sim 17\%$ increase in the second hyper-polarizability

results in less then 1 % change in $\chi^{(2)}$, indicating the stability of the calculated results with respect to the choice of field strength. Similarly, Table 6.6 shows that the second hyperpolarizability is also stable in a fourfold range of electric field. The eight-fold increase in electric field pushes the molecule out of the region of stability, but only by 17 %.

Table 6.7 Third order susceptibilities

Molecule	$\hbar\omega$	$\chi_{||}^{(3)}$	Experiment	GF
Ar	1.175	2354	1000±100	2283
Kr	1.175	5312	2790± 270	5064
Ne	1.175	191	79± 8	189
N_2	1.790	1440	1295± 206	1663
C_6H_6	1.790	59141	23810± 460	58500

The energy, $\hbar\omega$, is in eV, and $\chi_{||}^{(3)}(-3\omega; \omega, \omega)$ is in atomic units. The column GF shows the results obtained by the method of Iwata and Yabana [1] using the same ground state orbitals as in our calculations, and c) denotes the result of Ref. [1]

6.2 Comparison to Experiment and DFPT Calculations

In Tables 6.4 and 6.7 our results are compared with experiments and other calculations. The calculated results are close to the results obtained by using the method of [1], and the results of Salek et al. [7] and Andrande et al. [8]. On the experimental side, the second order susceptibilities of the CO, H_2O and HF molecules show a rather good agreement with the measurement. On the other hand, the second order susceptibilities for H_2S, NH_3, and majority of third order susceptibilities overestimate the experimental data by the factor of three. We expect that more sophisticated exchange-correlation functionals, such as B3LYP and LB94 will improve agreement with experiment. The discrepancy between the theory and experiment is due to several factors. The most important ones are (a) absence of nuclear motion, (b) condensed phase effects and (c) traditional shortcomings of the LDA functionals. For the CO and H_2O molecules we calculated the dispersion curves that demonstrate the correct qualitative behavior in a non-resonant spectral sectors (see Fig. 6.1). The figure shows the results of calculations done by others [7, 8] as well as experimental results [6] and the results obtained using method of Iwata and Yabana [1]. While none of the theoretical results matches experimental data for H_2O molecule, all calculations for CO molecule show better agreement. The discrepancy with experiment strongly depends on the level of theory and less on the method of calculations. Hatree-Fock underestimates CO experimental data and overestimates H_2O data. In contrast, TDDFT and DFPT calculations overestimate CO data and underestimate H_2O data. When the same ground state is used, difference between real-time TDDFT calculations and DFPT calculations using Iwata and Yabana algorithm are close to each other then calculations within the DFPT by others [7]. At the same time calculations within modified Sternheimer approach of Andrade et al. [8] for H_2O molecule are close to ours. One may conclude that:

- Level of the theory plays the decisive role in determining realism of calculations.
- ALDA fares better then Hartree-Fock, but the differences depend on specifics of molecular structure.
- ALDA functional is the key source of discrepancy between our calculations and experiment.

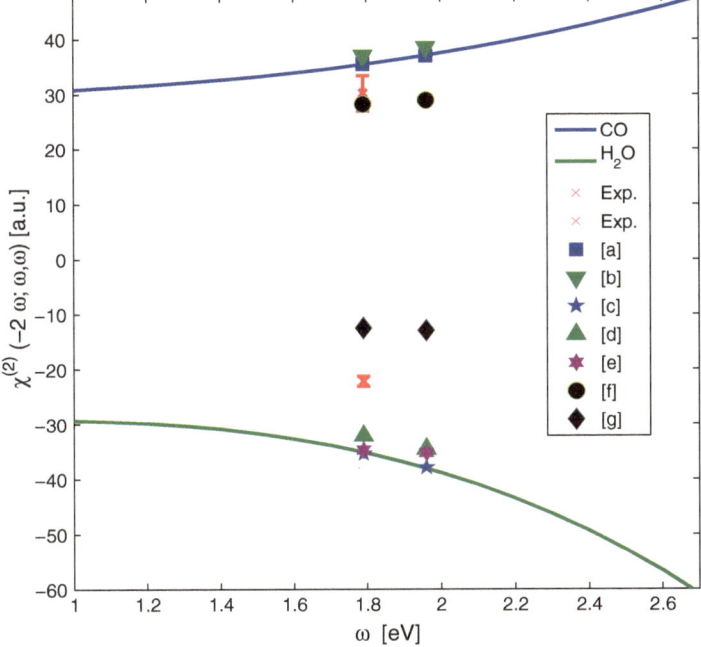

Fig. 6.1 Dispersion curves for the second order nonlinear susceptibility $\chi^{(2)}(-2\omega; \omega, \omega)$ of CO (*top, blue color*), and H_2O (*bottom, green color*) molecules. The experimental data is from [6]; **a, c** denotes results obtained by using the method of Iwata and Yabana [1]; **b, d** shows the results of Ref. [7], **e** denotes the results from [8], and **f, g** are Hartree-Fock calculations from [7]

One should keep in mind that the above observations are drawn from data for small and medium organic molecules calculated under Kleinman symmetry conditions[3] and may have limited generality.

6.3 Silver Dimer

As the last example we explore the nonlinear response of Ag_2 cluster. The silver dimer is one of a range of silver clusters up to Ag_{32} that was studied recently [10]. 2.612 Å value is used for the distance between the two atoms [11]. Twenty-two valence electrons were used in calculations. The dimer was enclosed in 14 Å cube with mesh step 0.25 Å. PZ exchange-correlation functional was used. The excitation frequency of external field was set to $\hbar\omega = 1.17\,eV$. Three fields were used: $\lambda = \{0.013, 0.025, 0.05\}$ (V/Å) The second hyperpolarizabilities were extracted using method

---

[3] See appendix B for definition.

Table 6.8 Spatially averaged second hyperpolarizabilities $\chi_{\parallel}^{(3)}$ of silver dimer at $\hbar\omega = 1.17\,\text{eV}$

Cluster	$\chi_{\parallel}^{(3)}(-\omega)$	$\chi_{\parallel}^{(3)}(-3\omega)$
Ag$_2$	$8.8 \times 10^4 - 5.3 \times 10^3\,\text{i}$	$-1.3 \times 10^5 - 8.3 \times 10^5\,\text{i}$

Data is in atomic units

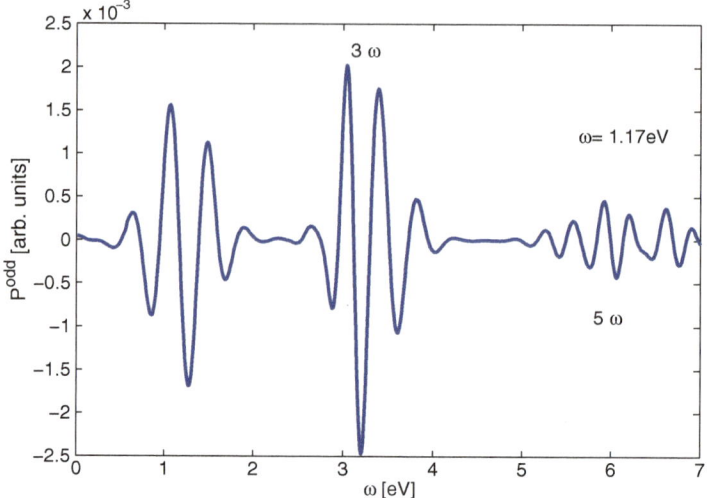

Fig. 6.2 Odd order nonlinear optical response of Ag$_2$ cluster

II described in Chap. 5. The averaged second hyperpolarizabilities are presented in Table 6.8, and the characteristic third order nonlinear response is shown in Fig. 6.2.

Considering the fact that it is a two atom molecule, nonlinear response is very strong, with absolute value of third order susceptibilities $\sim\!10^4$ a.u.

Figure 6.2 shows that nonlinear response at ω_0 is mostly regular,[4] and is practically real. Nonlinear response at $3\omega_0$ is irregular, it has a contribution from fifth order and a large imaginary part. Also, fifth order response clearly shows up at $5\omega_0$. At least five different amplitudes of external field are needed to resolve the contributions from fifth order. Because we used three, it makes susceptibility values ambiguous at $3\omega_0$. Alternatively one may try to use smaller values of the external field in order to reduce the fifth order contribution.

This example shows that the real-time approach gives more information about the nonlinear response than one would expect from a single frequency excitation. If a perturbative method such as Sternheimer gives just a number representing hyperpolarizability at certain frequency, the real-time method calculates and shows behavior of the nonlinear response within a finite interval. This data allows to graphically ascertain the sign of the hyperpolarizability as well as whether it has a substantial imaginary part without even extracting the hyperpolarizabilities. It also provides

---

[4] Here it means that the nonlinear response largely follows the cube of the external field.

information on how many orders of the response are being excited and if the number of propagations is adequate.

6.4 Summary

In summary, we have explored the applicability and reliability of real-time real-space TDDFT method for calculations of molecular hyperpolarizabilities. The calculations are stable with respect to the variation of field strength, and there is no convergence problem associated with the basis functions. The calculations are in line with the results obtained by other methods using LDA functional and represent accurate estimates of nonlinear optical properties at the level of TDDFT.

The real-time TDDFT (RT-TDDFT) method to calculate response functions is fundamentally different from other perturbation theory based methods (such as the modified Sternheimer approach [1, 8]). The density and all observables that are derived from it are obtained from the single particle states that are explicitly time dependent and non-perturbative. The inclusion of nuclear motion is made simple in this case (for example by using Ehrenfest-type nuclear dynamics [12]) and does not require any changes in extraction algorithms. There are also critical computational differences between real-time and the perturbative methods based on the Sternheimer approach. The modified Sternheimer method relies on linear solvers and their performance determines the quality of the derived response functions. For large, complex molecules the convergence of these algorithms becomes problematic even at off-resonant frequencies. In near resonance they stop working even for the small molecules. The real-time propagation is stable near the resonance, and produces the data with the same efficiency. The real-time methods are computationally demanding, but these demands are predictable and propagation routines are easily scalable. On another hand it is not possible to say how many iterations will take to achieve convergence for a linear solver.

References

1. J.-I. Iwata, K. Yabana, G.F. Bertsch, J. Chem. Phys. **115**, 8773 (2001)
2. J.P. Perdew, A. Zunger, Phys. Rev. B **23**, 5048 (1981)
3. S.H. Vosko, L. Wilk, M. Nusair, Can. J. Phys. **58**, 1200 (1980)
4. R. van Leeuwen, E.J. Baerends, Phys. Rev. A **49**, 2421 (1994)
5. K. Yabana, T. Sugiyama, Y. Shinohara, T. Otobe, G.F. Bertsch, Phys. Rev. B **85**, 045134 (2012)
6. D.P. Shelton, J.E. Rice, Chem. Rev. **94**, 3 (1994)
7. P. Salek et al., Mol. Phys. **103**, 439 (2005)
8. X. Andrade, S. Botti, M.A.L. Marques, A. Rubio, J. Chem. Phys. **126**, 184106 (2007)
9. Y. Takimoto, PhD Thesis, University of Washington, 2008
10. V. A. Goncharov, PhD Thesis, Vanderbilt University, 2014
11. V. BonacicKoutecky, L. Cespiva, P. Fantucci, J. Koutecky, J. Chem. Phys. **98**, 7981 (1993)
12. J.L. Alonso et al., Phys. Rev. Lett. **101**, 096403 (2008)

Chapter 7
Extension to Condensed Matter and Outlook

7.1 Calculations in Dense Media

The last issue that we want to address before concluding this brief is extension of
the real-time method to the condensed matter. So far all calculations were done for
a single molecular structure in vacuum. The case of highly diluted gaseous media
would simply require to multiply the calculated hyperpolarizabilities by the number
of molecules per unit volume N: $\chi^{(n)} \rightarrow N\chi^{(n)}$. When the density of molecules
becomes high, the correction is generalized by including additional multiplicative
local field factor $\mathscr{L}(\omega)$ that accounts for dipole-dipole screening in dense media:

$$\mathscr{L}(\omega) = \frac{1}{1 - \frac{4\pi}{3} N\alpha(\omega)}, \tag{7.1}$$

where $\alpha(\omega)$ is polarizability. The corrected susceptibilities become:

$$\chi^{(n)}(\omega) \rightarrow N(\mathscr{L}(\omega))^{n+1}\chi^{(n)}(\omega). \tag{7.2}$$

The local factors can be used to correct a variety of disordered media. Periodic solids
that include such important class as semiconductors require a different approach.

7.2 Calculations in Case of Periodic Solids

The application of the real time method to periodic solids encounters two difficul-
ties. One is that the usual definition of polarization through position operator $\hat{r}$ is
ambiguous [2–4]. The other that external perturbation taken in dipole approxima-
tion as a scalar field $-\mathbf{r} \cdot \mathbf{E}(t)$ violates periodicity of crystal field and therefore cannot
be used in such form [5]. Both of these problems are addressed simultaneously by

© The Author(s) 2014
65
V. Goncharov, *Non-Linear Optical Response in Atoms, Molecules and Clusters*,
SpringerBriefs in Electrical and Magnetic Properties of Atoms, Molecules, and Clusters,
DOI 10.1007/978-3-319-08320-9_7

amending the real-time propagation with Maxwell-Schrödinger formalism developed by Bertsch, Yabana et al. [1, 5, 6].

In this theory, the external field is represented by a vector potential $\mathbf{A}$:

$$\mathbf{E} = -\frac{\partial \mathbf{A}}{\partial t}, \tag{7.3}$$

$$\nabla \cdot \mathbf{A} = 0. \tag{7.4}$$

This eliminates the external scalar field and the associated translational symmetry violation. The electron-photon dynamics is described by a coupled Maxwell-Schrödinger system of equations. The coupling variable is the induced current density $\mathbf{j}_{ind}$. The polarization is calculated form the spatially averaged current density:

$$\mathbf{P}(t) = \frac{\int_{t_0}^{t} \langle \mathbf{j}_{ind}(t') \rangle \, dt'}{\Omega}, \tag{7.5}$$

where Ω is the volume of the unit cell. The theory also includes the dynamic screening, thus no additional local field factors are necessary. This methodology was successfully applied to the calculations of third order susceptibilities in carbon diamond and crystal silicon [7].

7.3 Future Directions

In the preceding chapter we have already discussed the advantages of the real time method that come from its ability to calculate both off- and near- resonance susceptibilities for practically any molecular system, as well as its computational predictability that makes automation of large scale calculations simple. Among the biggest issues that limit accuracy of the TDDFT based calculations we pointed out the inherent limitations of the available exchange-correlation potentials. It is expected that new generations of exchange—correlation potentials will cut the discrepancy with experimental data to few percents from today's typical errors of 50 % or larger. The last issue that we want to mention is related to the real space implementation of the method. Use of the uniform grids has two advantages: simplicity and efficiency in implementation and insensitivity toward specific molecular structure as opposed to basis set biases in calculations that use basis sets. Hovever, the efficiency of the uniform grids is limited by the need to have fine mesh for accurate calculations. Because the fine mesh covers uniformly entire computationl cell, there is an unavoidable inefficiency. The obvious solution may be to use non-uniform adaptive mesh. Implementation of the adaptive grids for real time calculations would be an important step in increasing the efficiency and ultimately the accuracy of the method.

References

1. K. Yabana, T. Sugiyama, Y. Shinohara, T. Otobe, G.F. Bertsch, Phys. Rev. B **85**, 045134 (2012)
2. A. Dal Corso, F. Mauri., Phys. Rev. B **50**, 5756 (1994)
3. A.D. Corso, F. Mauri, A. Rubio, Phys. Rev. B **53**, 15638 (1996)
4. R.D. King-Smith, D. Vanderbilt, Phys. Rev. B **47**, 1651 (1993)
5. G.F. Bertsch, J.-I. Iwata, A. Rubio, K. Yabana, Phys. Rev. B **62**, 7998 (2000)
6. Y. Shinohara et al., Phys. Rev. B **82**, 155110 (2010)
7. V.A. Goncharov, J. Chem. Phys. **139**, 084104 (2013)

Index

© The Author(s) 2014
V. Goncharov, *Non-Linear Optical Response in Atoms, Molecules and Clusters*,
SpringerBriefs in Electrical and Magnetic Properties of Atoms, Molecules, and Clusters,
DOI 10.1007/978-3-319-08320-9